A Practical Guide To SharePoint 2013

Saifullah Shafiq

A Practical Guide To SharePoint 2013

Published by Wali Systems, Inc.

Copyright © 2014 by Saifullah Shafiq

ISBN-13: 978-0-9915203-0-5
ISBN-10: 0991520300

Library of Congress Control Number: 2014902311

Printed and bound in the United States of America.

Wali Systems books are available through booksellers and distributors worldwide. For further information about international editions, contact Wali Systems Inc directly at phone (301) 401-7291 or fax (810) 592-7291. Visit our Web site at walisystemsinc.com/press. Send comments to info@walisystemsinc.com.

All rights reserved. No part of this publication may be reproduced, stored in a retrieval system, or transmitted, in any form, or by any means (electronic, mechanical, photocopying, recording or otherwise) without the prior written consent of the publisher/author.

While every precaution has been taken in the preparation of this book, the publisher and author assume no responsibility for errors or omissions, or for damages resulting from the use of the information contained herein. This book is presented solely for educational purposes. The information contained in this book is provided without any express, statutory, or implied warranties. No warranty may be created or extended by sales representatives or written sales materials. Neither author, nor the publisher, or resellers, or distributors will be held liable for any damages caused or alleged to be caused either directly or indirectly by this book.

First Published: February 2014

Trademarks: Microsoft and SharePoint are registered trademarks of Microsoft Corporation in the United States and/or other countries. Facebook is a registered trademark of Facebook, Inc. Twitter is a registered trademark of Twitter, Inc. All other trademarks are the property of their respective owners.

Cover: Wali Systems Design Services

I dedicate this book to the SharePoint community that is one of the best technical communities. SharePoint community has played a major role in establishing SharePoint as the market leader in portal technology.

Saifullah Shafiq

About the Author

SAIFULLAH SHAFIQ is a SharePoint consultant and author. He has been working with SharePoint since its first release. He started his career as a web developer 14 years ago. He has been contributing to the IT communities since 1999. His work (technical articles) has been published on hundreds of programming related web sites. He has also written for different IT magazines. This is Saifullah's fifth book on SharePoint.

Saifullah is a MCP (C#), MCTS (SharePoint) and MVP (SharePoint). Saifullah runs a SharePoint user group that has over 450 members.

Saifullah owns a development firm called Wali Systems, Inc. (http://walisystemsinc.com). Wali Systems specializes in SharePoint and InfoPath development. Wali Systems develops custom SharePoint applications and products. Wali Systems also provides SharePoint staffing, recruiting, consulting and support services.

Recently, Saifullah launched a SharePoint jobs portal called Hot SharePoint Jobs (http://hotsharepointjobs.com) that uses latest social media recruiting tools and is quickly gaining popularity.

Table of Contents

Introduction

Welcome to *A Practical Guide To SharePoint 2013*. SharePoint is a great product that helps teams stay connected. It helps teams to be more productive by providing a robust infrastructure. Users can easily access people, documents, and information that they need. It plays a major role in helping organizations to convert data into information. Before SharePoint, data used to reside in different silos. Some data would be on the local disks, some in the database, and some in external disconnected drives. With SharePoint, organizations can now store all data in one place and give access to the people who matter. SharePoint comes integrated with the popular Microsoft Office tools like Word, Excel, PowerPoint, etc.

SharePoint is a huge product. Each component or feature deserves a separate book. Discussing all great features in detail in one book is not possible. I have tried to include the necessary content that will help you get started straight away. For beginners, the main problem they face is to setup an environment for the SharePoint deployment. This book shows how to prepare the environment and then install each piece of software one by one. I started writing this book in late 2012 when the RTM version of SharePoint 2013 was released. I regret I did not publish the book earlier as I always wanted it to be perfect. During the past one year, several books have arrived on the market. This book is not competing with other books but it still has unique content that will help readers in learning the art of SharePoint deployment. Furthermore, to keep it interesting for the readers, I have added several new chapters that are focused on the latest features and hot technologies, for example, SharePoint online publishing.

The first half of the book shows how to install SharePoint 2013 on a clean machine. The intended audience is people who have never installed SharePoint before. The book guides users in a step-by-step format and explains everything needed to install the software. This includes preparing the machine for the installation. I have used Oracle VM VirtualBox to install the software. You can choose between other options but Oracle VM VirtualBox is free and meets our requirements.

Each step in the book shows a screenshot that will help users to better understand the step. Host computer used during the writing of this book was a

Windows 7 (64-bit) machine. Guest machine used was Oracle VM Virtualbox version 4.1.6 r74713 with Windows Server 2008 Service Pack 2 (64-bit).

Who Should Read This Book

This book offers a comprehensive guide on SharePoint deployment. So the first part of the book is more focused towards the SharePoint infrastructure and deployment and will help the administrators or the users looking to tread the path of becoming SharePoint administrators. Network folks who are not necessarily SharePoint administrators but want to learn how to install this cool product will also gain lot of useful information from this book. Second part of the book contains information on some of the popular features of SharePoint. It discusses creating web-enabled InfoPath forms, setting up SSL, setting up Extranets, etc. So, second part is for the power users and advanced users who are comfortable with using SharePoint.

Assumptions

A Practical Guide to SharePoint 2013 is divided into two parts. First half is designed for the readers who are just beginning to learn SharePoint and would like to install it locally on their machines or in a virtual environment. The book touches only briefly some of the topics that you will find covered in more detail elsewhere. This book will be helpful to server administrators, site owners, infrastructure folks, and developers. There is something in this book for every SharePoint user.

All code presented in the book has been tested for accuracy and correct functioning but since technology/applications/products change underlying code frequently, for example, Twitter APIs change from time to time, facebook updates its APIs, therefore, the code functionality presented in the book may slightly change by the time the book reaches your hands and you may have to slightly update the code to make it work. You are encouraged to subscribe for updates on the book's website so that if something changes, you can be notified.

How This Book Is Structured

This book is structured to build logically on the skills you learn as you progress through it. After the initial introduction and deployment chapters, the book moves into the more advanced part of the platform. Each chapter focuses on new feature. If you are an advanced user, you can skip the initial chapters and go directly to the topic that you are interested in.

What You Need To Use This Book

First, you will need a copy of SharePoint 2013 and that's what this book focuses on in the initial chapters. You will learn to install SharePoint 2013 in your local environment. Once you have it installed, you can play with the product and learn how to use it and how to implement advanced solutions.

Conventions

To help you get the most from the text and keep track of what's happening, I've used a number of conventions throughout the book.

Notes and tips are shown in the boxes like this one.

As for styles in the text:

- Important words are shown in **bold**.
- Source code is shown in a box (see below) and is labeled as **Listing**.

```
<system.web>
  <membership>
    <providers>
    </providers>
```

```
    </membership>
  </system.web>
```

- Images are labeled as **Figure** and are numbered.

Source Code

All the source code used in the book is available for download at http://practicalguidetosharepoint.com.

Chapter 1

SharePoint Deployment

In this chapter:

- SharePoint Deployment Environment Preparation
- SharePoint Pre-requisites
- SharePoint Installation

Brief Introduction

SharePoint is one of the most successful products ever launched by Microsoft. It has been popular amongst enterprises because they take advantage of its rich features. On the other hand it is equally popular amongst consultants, contractors, and geeks. SharePoint started gaining immense popularity back in 2004. That was the time when SharePoint really started picking up. It has been on the market since 2001 though. Many people switched to SharePoint back then. It was easier for the .NET developers to understand the functionality and they were the ones who delved into SharePoint development first. SharePoint started as a document sharing system but after it became popular with the companies, it was converted into a full-fledged document management system. It also gained popularity because of its team collaboration features. It was easy to use and gelled easily in to the existing infrastructure that companies already had. I was a PHP programmer before I jumped on the SharePoint band wagon and I must admit there has been no looking back since then. Before PHP I used to develop web applications using classic ASP.

With the latest version SharePoint 2013 Microsoft has offered so many rich features that you don't have to look around for small applications. It's a complete suite of applications that you can use in your organization. Microsoft earned approx. $1.3 billion in revenue in 2010. It was the hottest Microsoft product in 2011 and if you ask me, I believe it will still be hot in 2014. There is something for everyone in SharePoint 2013. Some of the most popular features in SharePoint 2013 are as following:

1. SharePoint Business Connectivity Services
2. SharePoint Content Management Features
3. Social Computing Features
4. Business Intelligence
5. Enterprise Search
6. Integration with enterprise social features (Yammer)

Full list of SharePoint 2013 features can be seen on the following page:

http://sharepoint.microsoft.com/en-us/product/Pages/Features.aspx

In early 2011, Mark Miller published SharePoint 2010 adoption survey results. Mark is a popular community builder and participates actively in the creation of global online communities and is a co-founder of NothingButSharePoint.com. I was present at that event where survey was conducted by Global 360 and EndUserSharePoint.com (also owned by Mark). According to the survey results, SharePoint 2010 adoption surged in 2010. Another important result was that SharePoint 2010 had solidified its position as mission critical part of IT and business.

Other industry experts also conducted surveys and all results pointed more or less to the same facts that SharePoint's popularity was on the rise. As mentioned above, SharePoint consulting firms, consultants, contractors, everyone took advantage of this gold mine called SharePoint. Especially companies that were Microsoft partners made new clients throughout all these years when SharePoint was becoming popular. According to some statistics, 80% of the businesses are expected to be using SharePoint. Can you believe it? SharePoint is not only for big companies any more. Small and medium businesses are also taking advantage. I have implemented SharePoint in retail businesses. Small retailers including gas and service stations are taking advantage of SharePoint. Throughout 2013 I had been setting up small business sites in the cloud using SharePoint Online for small clients and all clients are happy with their sites. It has made managing their businesses super easy. They use it in conjunction with Microsoft Office that they were already using.

You can judge SharePoint's popularity from the fact that there is a serious shortage of good resources on the market. Companies are finding it hard to find good resources. SharePoint is so hot that I have seen people advising other people who are jobless to start learning SharePoint. I was recently conducting interviews for a client. We were interviewing SharePoint resources (developers) and I noticed that 60% of the candidates had never developed a SharePoint application. They sent their resume because they installed it last year on their home PC and they were so confident of finding a job that they sent their resume even though advertised position clearly said "SharePoint Developers with at least 2 years' experience". We needed far more experienced resources but because we knew the market situation, we reduced the experience requirement to 2 years and still found it difficult to find good experienced resources. If you went to tech job sites and searched for "SharePoint" keyword, you'd find thousands of jobs listed. Usually job market becomes sluggish in December because of the festive season but SharePoint hiring is continuing despite the cold chilly weather, negative comments everywhere regarding the economy and the holidays in the final days of December (I am writing this section in December and the book will be available in stores next month, that is, Jan 2014).

Training companies also made big money. Organizations needed trainers on site to train end users. Demand for SharePoint training was such that online trainers also jumped in and made money by providing training online. Online training is not new but if you look at the number of companies that provided online SharePoint trainings in 2013, you'd be amazed.

Ask companies that make products for SharePoint! How their profits surged by selling SharePoint products! Selling is an art. If you know how to sell and if you are selling SharePoint products, the combination is deadly. You are destined to make good money. I know several small vendors that include personal friends as well. They made fortunes by selling SharePoint web parts to clients in the East coast, West coast and Mid-West. Not only the local market, but the demand for SharePoint products is huge in the international markets as well. Just imagine, if small vendors could make money just by selling web parts, think about the big vendors who sell full SharePoint application suites.

Good news is that SharePoint is still popular. Companies need resources. Contractors and consultants need more money. People working with SharePoint were happy in 2013. The way things are at the moment, it is apparent that 2014 will bring more joy on the faces of SharePoint consultants and contractors.

I know some people will scream about the little SharePoint annoyances that irritate the developer community but facts are facts. In spite of those annoyances and shortcomings, we still love SharePoint. To the world, SharePoint is a great product, it's software for the enterprises but for me, it's my livelihood. I live and breathe SharePoint.

Prepare your machine

Is your machine 64-bit compatible? Download following standalone utility from vmware.com that will check if your processor is 64-bit compatible or not:

http://download3.vmware.com/software/wkst/VMware-guest64check-6.0.0-45731.exe

If your machine is 64-bit compatible then second step is to confirm whether Virtualization Technology (VT) is enabled or not. To enable 64-bit support for virtual machines in VMware, you must enable Virtualization Technology (VT). VT provides greater flexibility and maximum system utilization. With VT, your machine will make the most of its underlying hardware. You can enable VT in the BIOS by going to advanced processor features and enabling Intel Virtualization Technology. The VT setting is locked at power on. After enabling VT in BIOS, save changes to the BIOS settings and then turn off power to the system. Your changes take effect when the system is turned on again. Simply rebooting the system without turning it off will not enable VT. This is called as power cycling the system.

Some laptops require a total power discharge to enable VT once you have set it in the BIOS. To do this, disconnect the external power, and remove the battery from the laptop. Then press the power button on the device to drain all power. Once this is done, replace the battery and reconnect the external power, and verify you are able to utilize VT.

Install VirtualBox

1. Install Oracle VirtualBox. If you want to use the version that I used when writing this book, you can download it from the following link:

   http://download.virtualbox.org/virtualbox/4.1.6/VirtualBox-4.1.6-74713-Win.exe

   If you want to download the latest build, download it from the following link:

   https://www.virtualbox.org/wiki/Downloads

   The latest build as of this time is 4.2 and that can be downloaded directly from the following link:

   http://download.virtualbox.org/virtualbox/4.2.0/VirtualBox-4.2.0-80737-Win.exe

2. After you have installed VirtualBox, enable **Shared Folders**.

a. From virtual machine settings or from menu, select **Shared Folders**.

b. You will see a green folder with a + sign on it on the right side. Click it to add a Shared Folder. In Virtual box, you can add shared folder as a machine folder or a transient folder. Add a machine folder.
c. Select path in **Folder Path**.
d. Give a name to the folder in **Folder Name**.
e. Check **Auto-mount** and **Make Permanent** options and click **OK**.
f. Click **OK** again.

Install Windows Server 2008 R2

3. Install Windows Server 2008 R2 (64-bit). I installed following version downloaded from MSDN:

en_windows_server_2008_r2_standard_enterprise_datacenter_and_web_with_sp1_x64_dvd_617601

SharePoint Requirements

4. Complete up to date list of hardware and software requirements for installing SharePoint 2013 is available at the following link:

   http://technet.microsoft.com/en-us/library/cc262485(v=office.15).aspx

   In a nutshell, 64-bit quad core machine with 2.5 GHz minimum per core is required. Microsoft recommends at least 8 GB RAM but you can also run it on 4 GB (for evaluation).

   Software requirements list is as following:

   - Windows Server 2008 SP 2 OR Windows Server 2008 R2
   - SQL Server 2012 (64-bit) or SQL Server 2008 R2 Service Pack 1 (64-bit)
   - Microsoft .NET Framework version 4.5 Release Candidate

   The setup program installs the following prerequisite for a single server with built-in database:

   - Microsoft SQL Server 2008 R2 SP1 – Express Edition

The SharePoint 2013 prerequisite installer (prerequisiteinstaller.exe) installs the following software, if it has not already been installed on the target server, in this order:

1. Microsoft .NET Framework version 4.5 Release Candidate (RC)
2. Windows Management Framework 3.0 Release Candidate (RC)
3. Application Server Role, Web Server (IIS) Role
4. Microsoft SQL Server 2008 R2 SP1 Native Client
5. Windows Identity Foundation (KB974405)
6. Microsoft Sync Framework Runtime v1.0 SP1 (x64)
7. Windows Identity Extensions
8. Microsoft Information Protection and Control Client

9. Microsoft WCF Data Services 5.0
10. Windows Server AppFabric
11. Cumulative Update Package 1 for Microsoft AppFabric 1.1 for Windows Server (KB 2671763)

Note: For SharePoint deployment exercise described in this book, I used Windows Server 2008 R2 and SQL Server 2008 R2 SP1.

Configure Active Directory

5. Download SharePoint 2013 RTM from MSDN subscribers download page.

6. After you have installed Windows Server 2008 R2, change the computer name to **SP2013** or whatever you want but it should be meaningful. I used **SP2013** and that's what you are going to see in the screenshots as well.

7. Disable Internet Explorer security. Go to **All Programs > Administrative Tools > Server Manager** and click **Configure IE ESC** link.

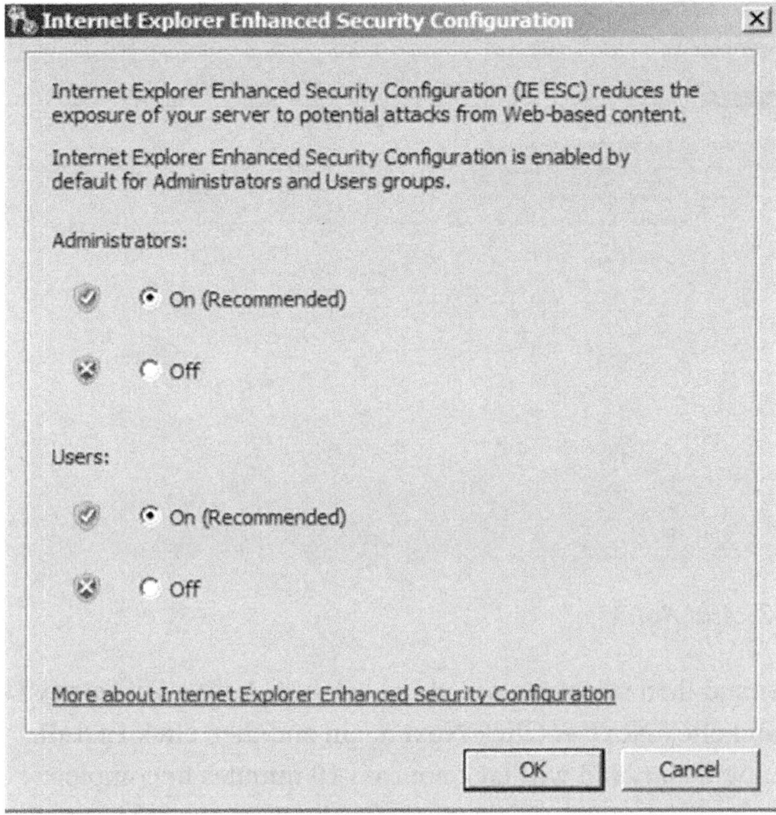

Figure 1-1: *Disable IE security*

Turn security off for both administrators and users and click **Ok**.

8. While still in Server Manager, click **Roles** and then click **Add Roles** link.

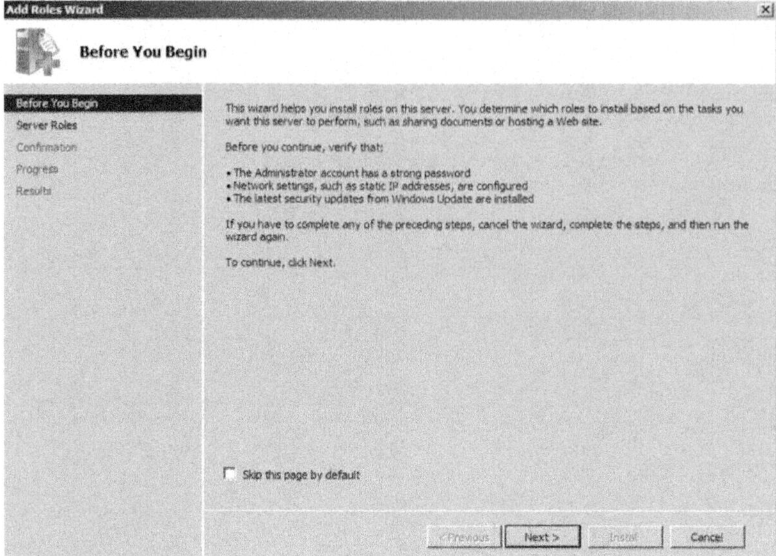

Figure 1-2: *Add Roles*

9. Click **Next** and then select the second check box **Active Directory Domain Services** and click **Next** >. Click **Next** again and then click **Install**. Installation will start and will take approx. 10 minutes to complete (depending on your computer's speed). Click **Close** once the installation has completed.

10. Click **Active Directory Domain Services** link.

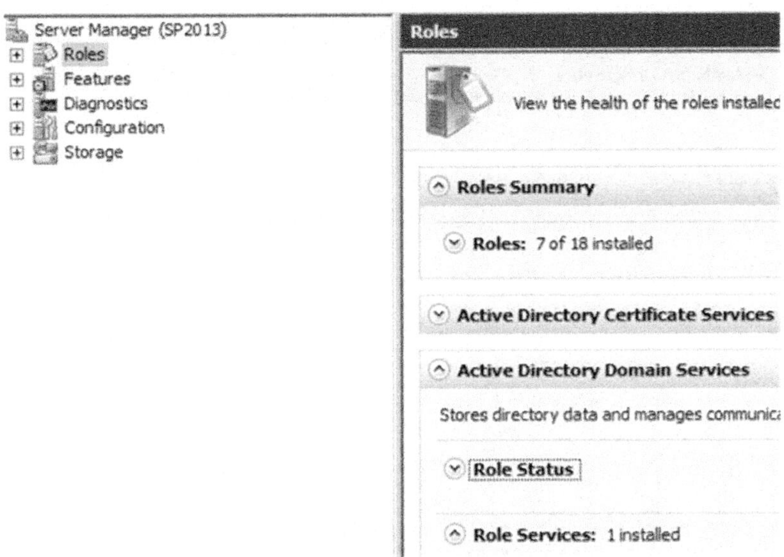

Figure 1-3: *Active Directory Domain Services*

You will see following warning message:

This server is not yet running as a domain controller. Run the Active Directory Domain Services Installation Wizard (dcpromo.exe).

Click the link in this message to install Active Directory Domain Services.

11. Click **Next**. Click **Next** again. Select the second option **Create a new domain in a new forest** and click **Next**.

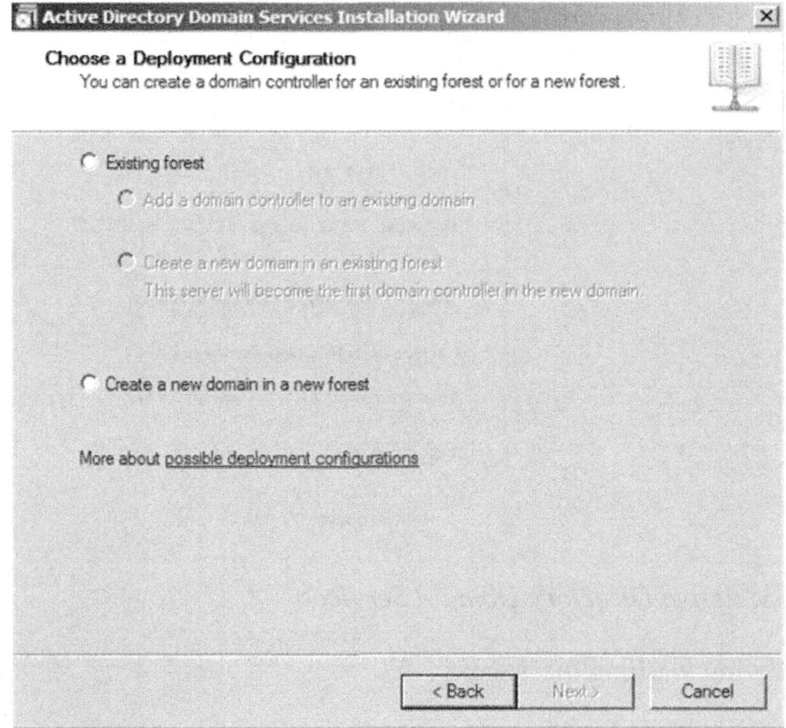

Figure 1-4: *Create a new domain in a new forest*

You will get following error if your current administrator password is set to blank.

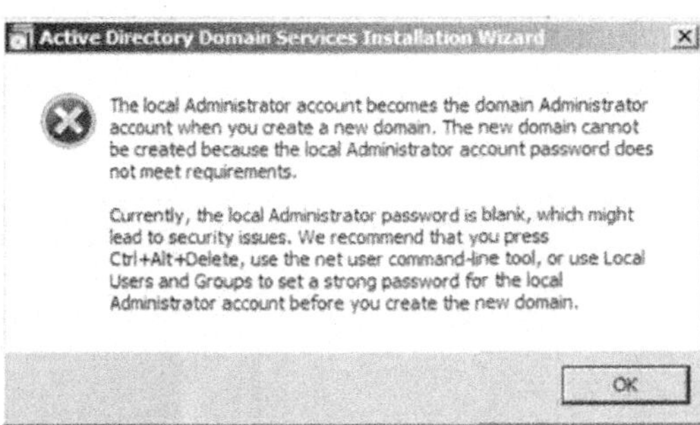

Figure 1-5: *Current admin password should not be blank*

Click **Ok** and change the Administrator password. Once you have changed the password, run the Active Directory Domain Services installation again (see steps above).

12. Provide a FQDN, for example, **sp2013.com** OR **server.sp2013.com** and click **Next**.

13. Select **Windows Server 2008** from the drop down and click **Next**.

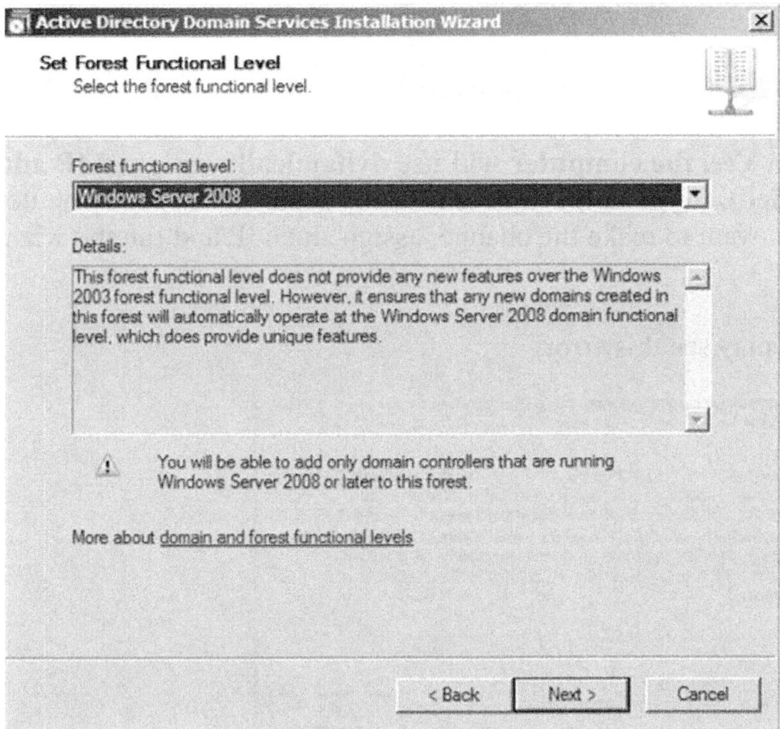

Figure 1-6: *Forest functional level*

14. Keep **DNS server** checked and click **Next**.

15. You will see following warning message if your computer has dynamically assigned IP address(es):

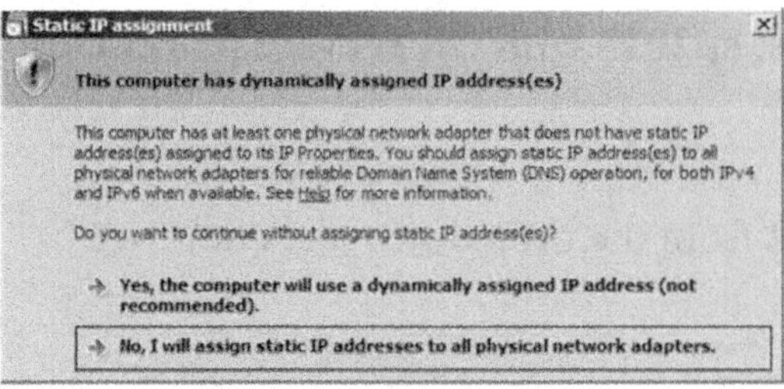

Figure 1-7: *This computer has dynamically assigned IP address(es)*

Select first option **Yes, the computer will use dynamically assigned IP address (not recommended).** If you don't want to assign a static IP to your computer and click **Next**. If you want to make the change, assign static IP and run the wizard again.

16. Next you may see this error:

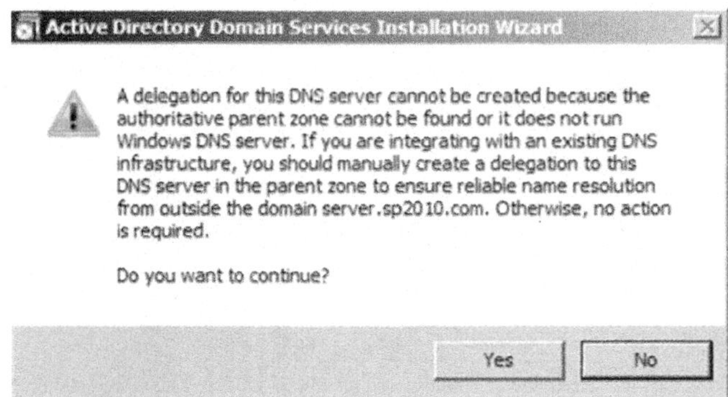

Figure 1-8: *A delegation for this DNS server cannot be created*

Click **Yes**.

17. You will be presented with the following screen:

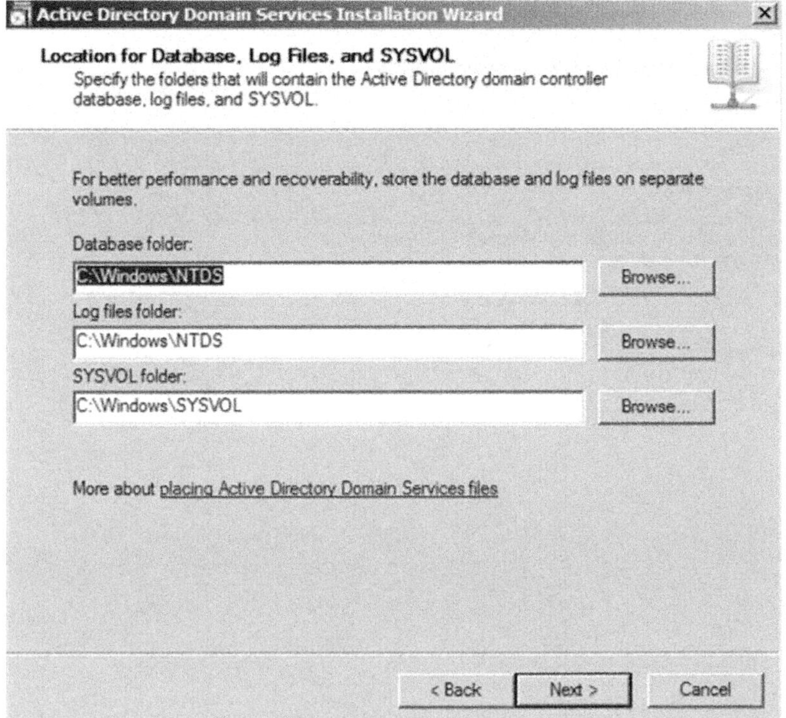

Figure 1-9: *Folders to store DC database, log files and SYSVOL*

If you have different partitions, save database and log files on different partitions. Select save locations and click **Next** (You can keep the default locations).

18. Enter password for Directory Services Restore Mode Administrator Mode and click **Next**.

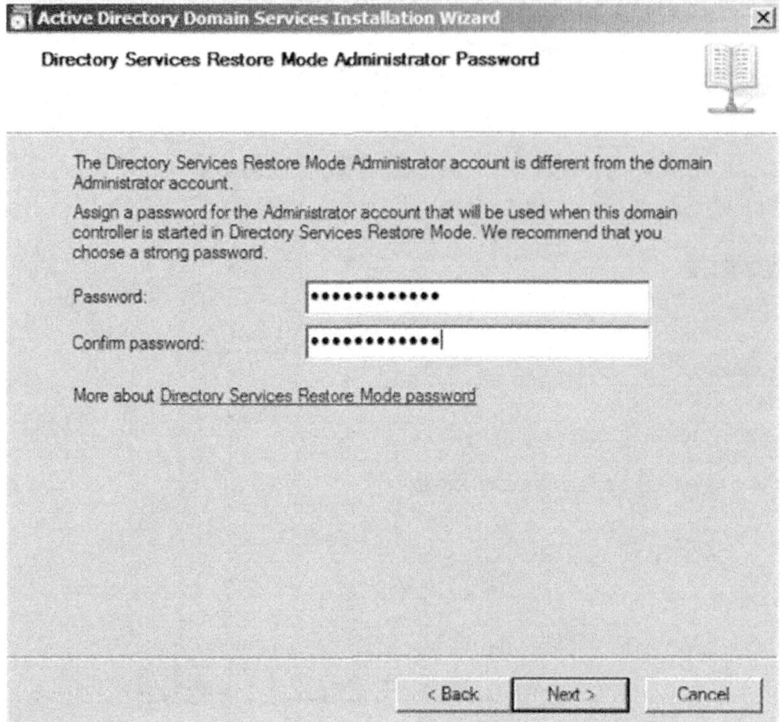

Figure 1-10: *Directory Services Restore Mode Administrator Password*

19. Review your selections and click **Next**. DNS Installation starts. Screen looks as following:

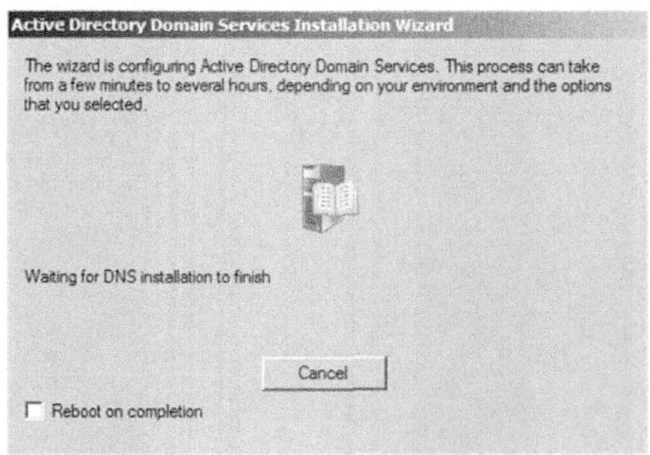

Figure 1-11: *Wizard is configuring Active Directory Domain Services*

The installation process will take a few minutes to complete.

20. Click **Finish** after the installation has completed.

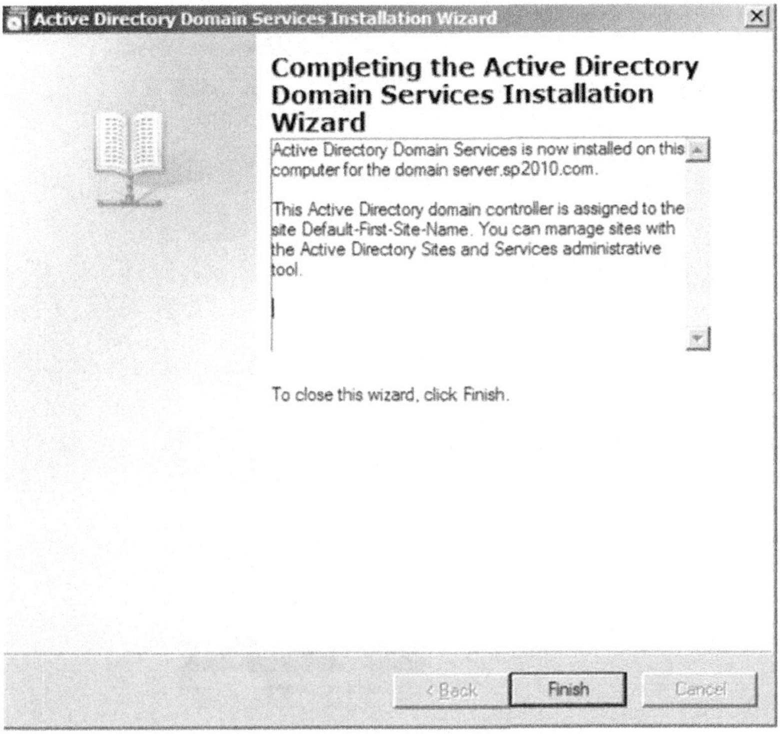

Figure 1-12: *Completing the Active Directory Domain Services Installation*

21. Click **Restart Now** button to restart your machine.

22. Open **Server Manager** and add a new user. This is for the SQL account. Do the following to add a new user:

- Expand **Roles** node
- Expand **Active Directory Users and Computers [FQDN]** node
- Expand [FQDN] node
- Right-click **Users** node, select **New** and then **User**

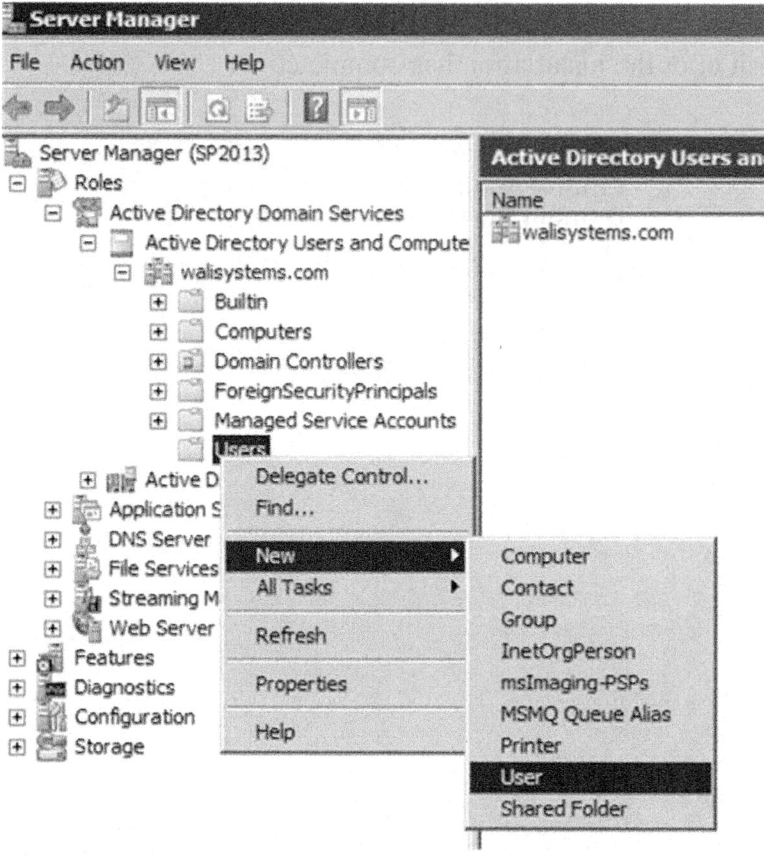

Figure 1-13: *Add a new user*

23. Enter **SQLAccount** in the **Full Name**. Enter **sqlaccount** in the **User logon name** box and click **Next**.

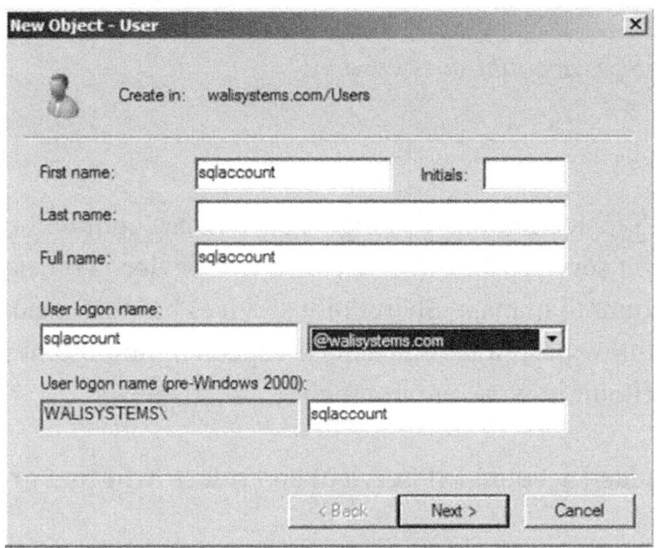

Figure 1-14: *Enter account details*

24. Enter new password in **Password** and **Confirm password** boxes. Uncheck all check boxes except **Password never expires** and click **Next**. Make sure you enter complex password, for example, you should not enter password same as user id, Enter one big alphabet and enter at least one digit. Make sure password length is not too short. If you don't follow these guidelines, you will get an error in the next step when you click the **Finish** button.

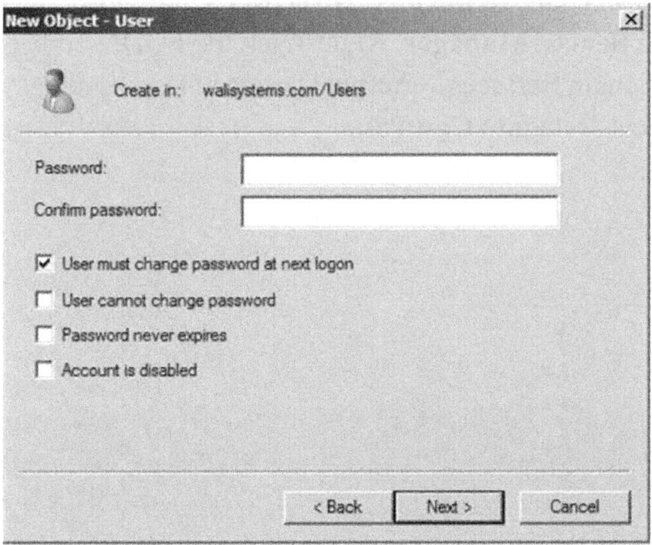

Figure 1-15: *Enter SQL account password*

25. Click **Finish**.

26. Repeat steps 22 – 25 to add another account. This new account will be used to manage SharePoint services and will be called **SPService**. You can also use your admin account to manage SharePoint services but that is not a good idea. It's better to follow best practices. This is especially important if you are deploying SharePoint in work environment.

27. Right-click newly added account (SPService) and select **Add to a group....**

28. Click **Advanced**.

29. Click **Find Now**.

30. Select **Administrators** and click **OK**.

31. If you decided to use your own admin account to manage SharePoint services then this step is not required. Just be sure that your admin account has full access to Active Directory. If you created a services account then you need to give this account **Replicated Directory Changes** permissions in Active Directory. In **Server Manager**, Right-click the FQDN node (Roles > Active Directory Domain Services > Active Directory Users and Computers) and select **Delegate Control...**.

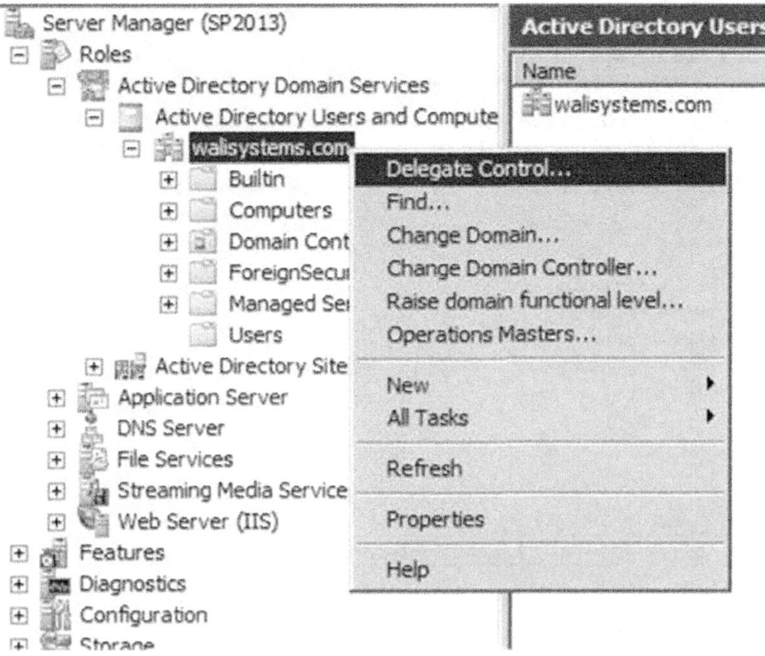

Figure 1-16: *Delegate control*

32. Click **Next**.
33. Click **Add**.
34. Enter the services account that you created above in the **Enter the object names to select box** and click **Check Names** button. If user name is resolved, click OK otherwise make sure you have entered the correct user.
35. Click **Next**.
36. Select the second radio button **Create a custom task to delegate** and click **Next**.
37. Keep the first radio button selected and click **Next**.

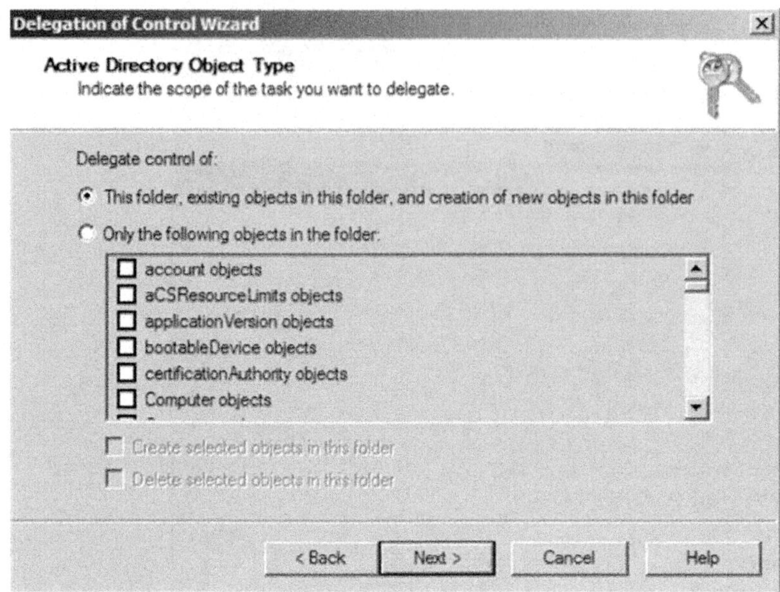

Figure 1-17: *Active Directory Object Type*

38. By default, first check box **General** is selected. Keep it selected and from permissions, check **Replicating directory changes** check box and click **Next**.

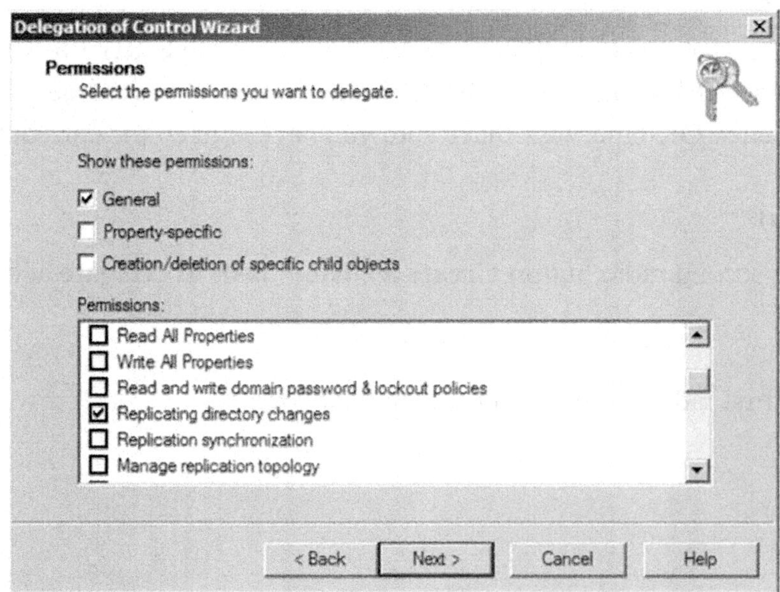

Figure 1-18: *Permissions you want to delegate*

39. Click **Finish**.

Summary

This chapter discussed installing VirtualBox and preparing the environment for SharePoint installation. Active Directory was setup was discussed in detail. Next chapter will show you how to connect your virtual environment to the Internet.

Chapter 2

Connect to the Internet

In this chapter:

- Find IP Addresses
- Configure Network Adapter Settings

How to connect your new virtual machine to the Internet

This is a straight forward process but sometimes it gets you entangled and all you do is waste your time trying to resolve the problem having no clue what to do and where to look for the solution.

First thing is to select the right network adapter in virtual machine settings. Select **NAT** in the network adapter box and then select correct adapter type.

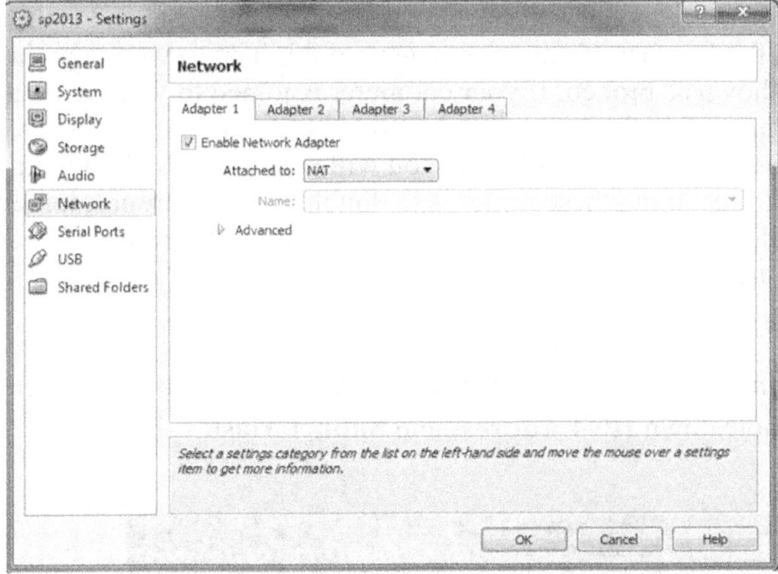

Figure 2-1: *Network Adapter settings*

You don't have to do anything in specific to configure the DNS but just in case if something goes wrong you can refer to this section to get help. For example, if you are unable to connect to the Internet after installing AD and DNS roles. Following are few tips that will help you get connected to the Internet.

1. Did you reboot your server after installing new roles? If not, the first thing you need to do is reboot your server.

2. Check your **Network and Sharing Center**. It will show you which part is malfunctioning.

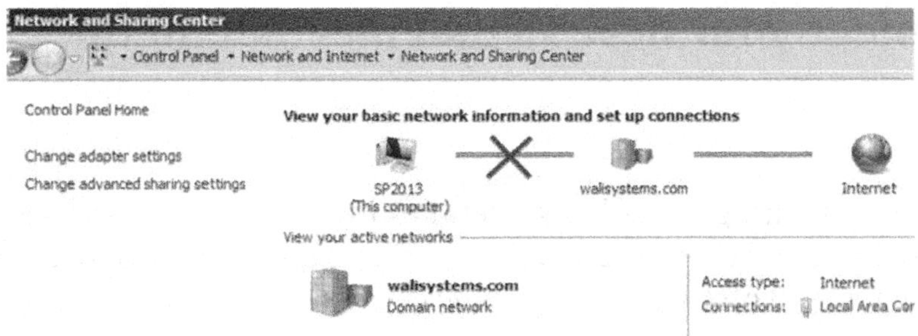

Figure 2-2: *Network and Sharing Center*

The above image has three main objects. One is the computer. Second is the domain name and third is the Internet. If your computer is not joined to your domain, the first link will be shown as broken. If your computer is joined to your domain but domain is not connected to the Internet, the second link will be shown as broken and in some cases, both links will be broken so you will need to fix both of them one by one. Easiest way of resolving these issues is to double click the troublesome link and Windows will resolve the issue automatically for you. Sometimes, Windows will fail in resolving it and you will be left with only one option and that is to set it up yourself.

3. Open command prompt and type **ipconfig /all**. It will show you some settings. Note down **IPv4 Address** and **Subnet Mask** values.

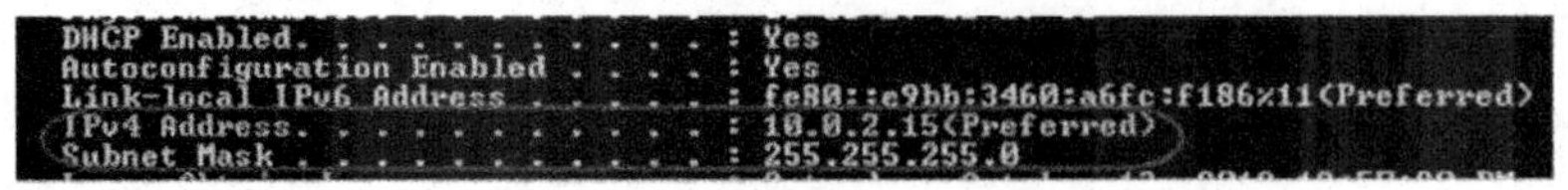

Figure 2-3: *IP settings*

4. In **Network and Sharing Center**, click **Change adapter settings**.

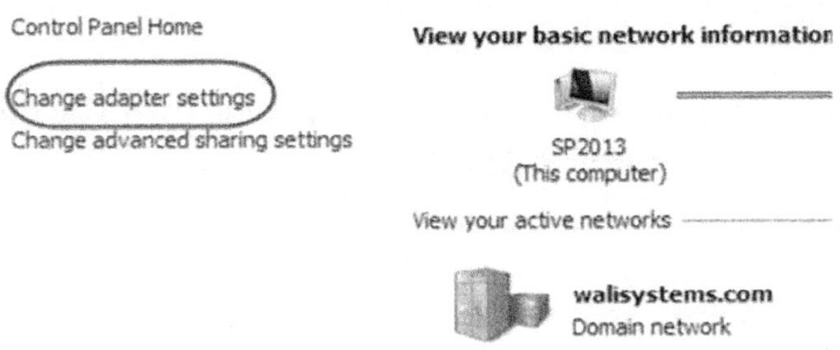

Figure 2-4: *Change adapter settings*

5. Right click **Local Area Connection** and select **Properties**.
6. Click **Internet Protocol Version 4 (TCP/IPv4)** and then click **Properties**.
7. Select **Use the following IP address** option.

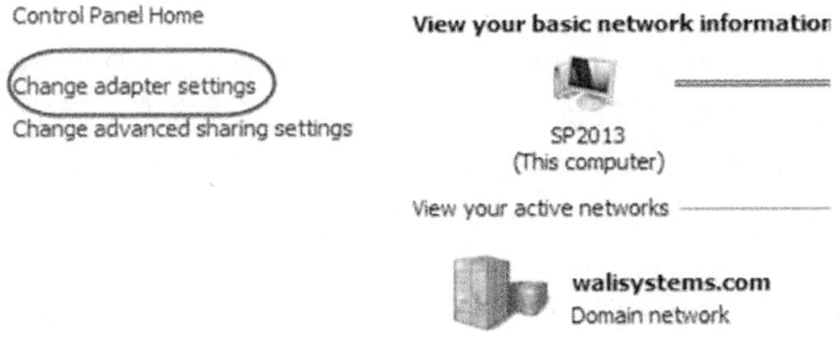

Figure 2-5: *Change adapter settings*

8. In **IP Address**, type the address of the client computer (copied from command prompt).

9. In **Subnet Mask**, type the subnet mask of the domain controller (copied from command prompt).

10. In **Default gateway**, type the address of the default gateway of the domain controller. If you don't know the address, type **10.0.0.1** for now.

11. Click **Use the following DNS server addresses**.

12. Now open a command prompt on your host OS and type **ipconfig /all** and then note down the values given in the **DNS Servers**. There will be two IPs, one for preferred DNS server and the other for Alternate DNS Server.

13. Come back to your guest OS (step 9) and enter values copied from host OS in the **Preferred DNS Server** and **Alternate DNS Server.**

14. Click **OK** and then **OK**.

15. Now go back to **Network and Sharing Center** and see if the problem has been fixed. If not, it will show you red X to indicate the problem. Continue with the following steps if the problem has not been fixed yet otherwise stop here.

16. Open **Server Manager.** Expand **DNS Server** node, expand **DNS** node, Expand **Forward Lookup Zones** and then click node that has your domain as its name. In the right side pane, you will see some entries that will have IP in the **Data** column. Delete these entries.

17. Go back to **Network and Sharing Center** and double-click the red X. This will start Windows diagnostics and will automatically fix the problem. Basically what Windows does is that it enables DHCP on the LAN. You can do that manually also but if for some reason Windows does not pick up the changes you make you can use Windows to make the changes for you. After Windows has enabled DHCP, your virtual machine will get connected to the Internet. I am assuming that you are using dynamically generated IP and not static one.

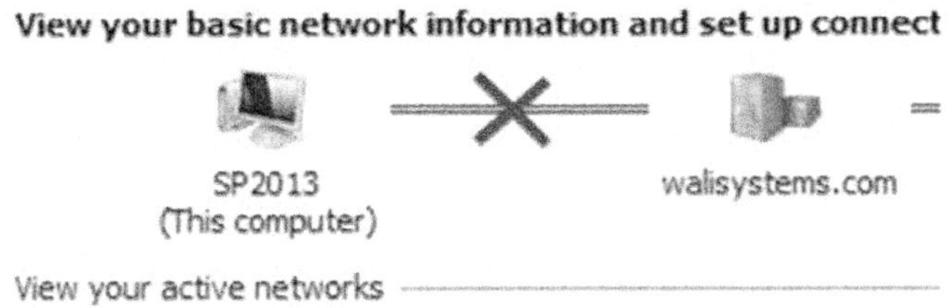

Figure 2-6: *View active networks*

To verify that DHCP has been enabled, open properties window of your LAN connection. Click **Internet Protocol Version 4 (TCP/IPv4)** and then click **Properties**.

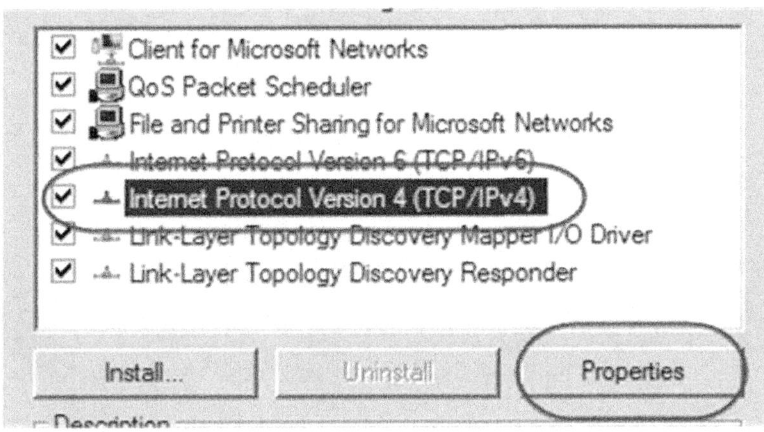

Figure 2-7: *IPv4 Properties*

Click **Advanced…** button. In **IP address** column, you will see **DHCP Enabled** listed.

Summary

In this chapter you learned how to configure your virtual environment to connect to the Internet. You saw how to change network adapter settings and where to look for the IP addresses.

Chapter 3

Install Pre-requisites

In this chapter:

- Install Pre-requisites
- Install .NET Framework
- Install SQL Server
- Install Application Server and IIS Roles

Install SQL Server Pre-requisites

Install .NET Framework 3.5 Service Pack 1

You need to install .NET Framework 3.5 SP1 before you can install SQL Server 2008 R2. If you miss this step, SQL Server setup will prompt you to install the pre-requisites.

You can download .NET Framework 3.5 SP1 from the following page:

http://www.microsoft.com/downloads/details.aspx?FamilyID=AB99342F-5D1A-413D-8319-81DA479AB0D7

It's an executable package. Package name is as following:

en_.net_framework_3.5_service_pack_1_x86_x64_ia64.exe

Copy this package to your virtual machine and run it. This will install .NET Framework 3.5 SP1 on your machine. Following is step-by-step installation detail:

1. Run the package and accept the license agreement and click **Next.**

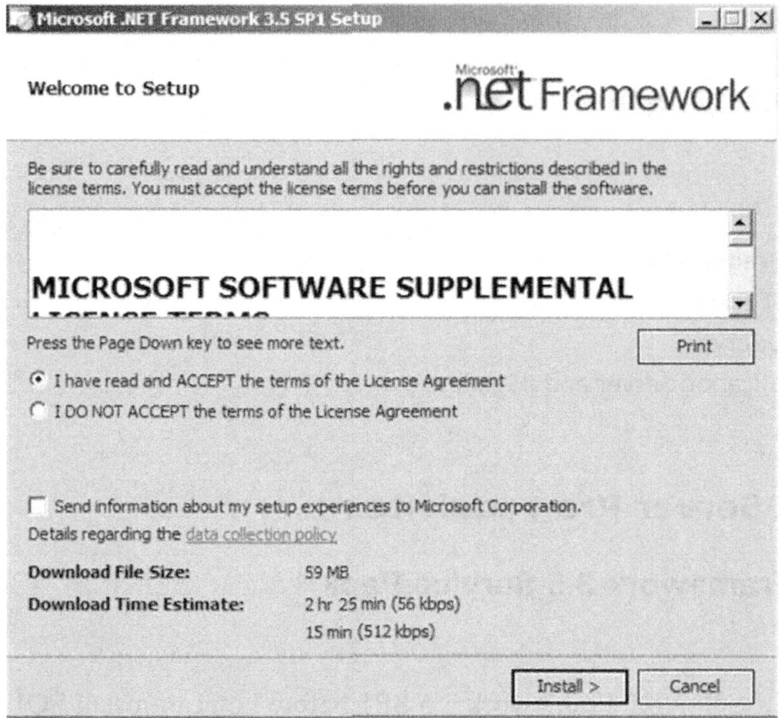

Figure 3-1: *Install Microsoft .NET Framework 3.5 SP1*

Installation will take at least 10-15 minutes. After the software has installed, click **Exit**. Click **Restart Now** button to restart your machine. This is important for the changes to take effect.

Please note that if you are using Windows Server 2008 R2 then you might be prompted to install **Application Server** role first before installing this package. To install **Application Server** role, open **Server Manager** and click **Roles**. Select **Add Roles** from the right side or right-click **Roles** and select **Add Roles.** Click **Next**, select **Application Server** and click **Install**. Wizard will guide you through the installation process. Complete installation steps are given in the later part of the book. See section **Install Application Server and Web Server (IIS) roles**.

Install SQL Server

You can copy the installer package to your virtual machine and run it from there or you can run it directly from the host machine. The installer is an ISO package, therefore, you can mount it and run directly from the host. If you haven't already downloaded SQL Server, you can download it from the Microsoft site. I will use SQL Server 2008 R2 SP1 for this walkthrough. If you haven't downloaded it yet, download the following package from Microsoft:

en_sql_server_2008_enterprise_x86_x64_ia64_dvd_x14-89207.iso (3,180,582 KB)

This is enterprise version for 64 bit machines.

Instead of unzipping the package as explained below, you can directly mount the iso image using the magicDisc software or any other mounting software and install the SQL server. MagicDisc is popular and reliable. It's easy to use as well. Download link is as follows:

http://www.magiciso.com/tutorials/miso-magicdisc-history.htm

To run it from your virtual machine, you will need to rename the .iso extension to .rar extension and then use WinRAR (http://rarlab.com) software to unzip the files to your hard disk. After unzipping the files, you can run the setup package directly from your virtual machine's hard disk. You can use other software also that can unzip the cab files (zipped files). There is some third party software that can read directly from the .iso image.

2. To mount the iso image, go to **VM** menu and select **Storage.**

3. Right-click **IDE Controller** and select **Add CD/DVD Device**.

4. Click **Choose disk** and browse to the .iso file.

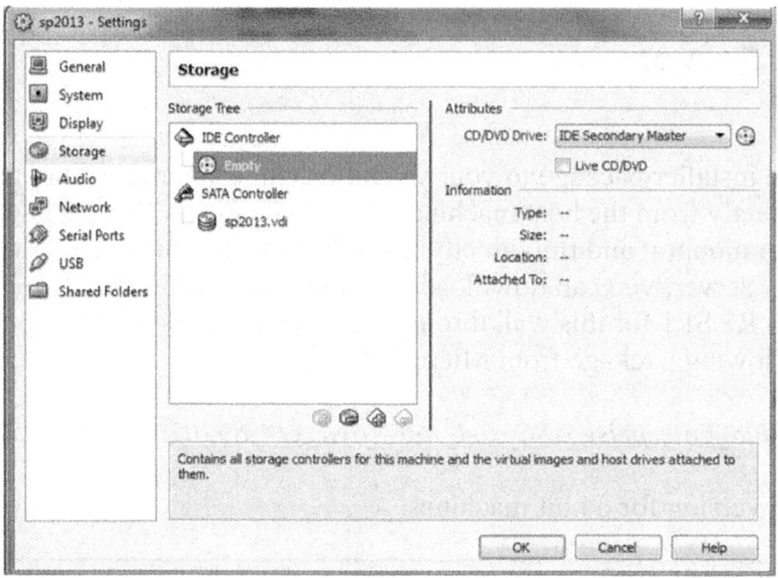

Figure 3-2: *Mount ISO Image*

5. Click **Open**.

6. Click **OK**.

7. ISO image has been mounted. Now click the DVD drive, AutoPlay options will be displayed. Double-click **Run SETUP.EXE** to start the installation.

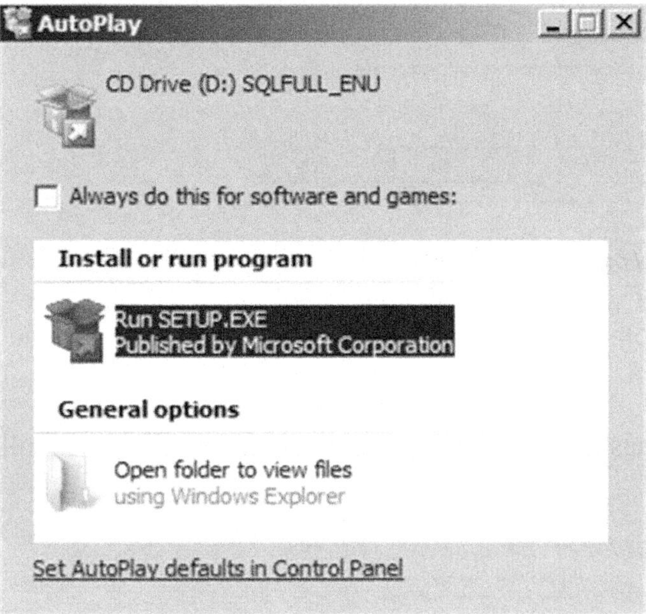

Figure 3-3: *Run SQL Server setup*

8. You may see the following alert:

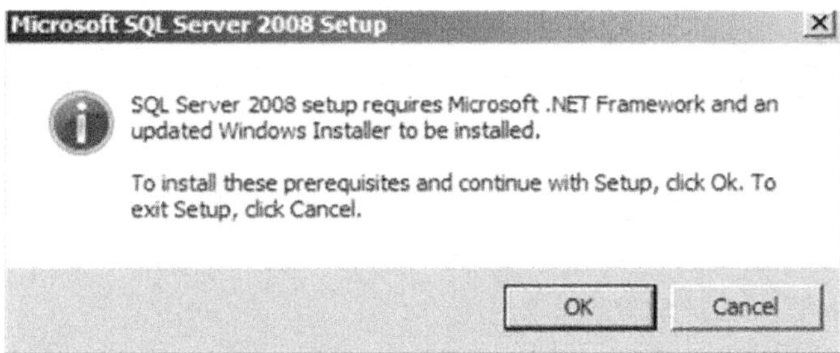

Figure 3-4: *SQL Server Setup*

The error occurs if Windows Installer 4.5 is not installed on your machine. Click **Ok**. You may be prompted to install a hotfix for Windows (KB942288). Click **Ok** to install it.

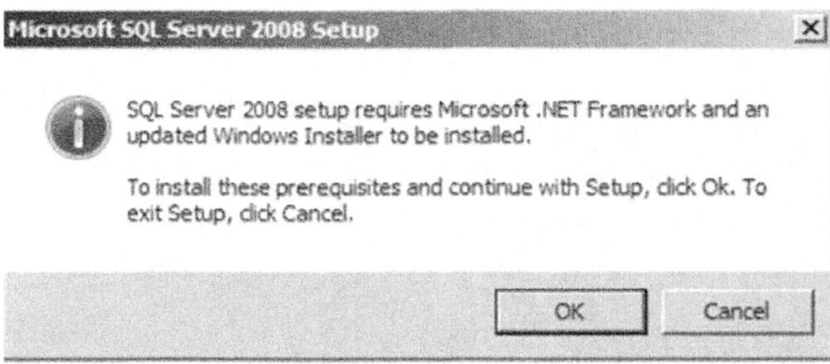

Figure 3-5: *Install Hotfix KB942288*

After the installation has completed, click **Restart Now** button to restart the machine.

9. After the machine has started, run SQL setup again. You will see following screen:

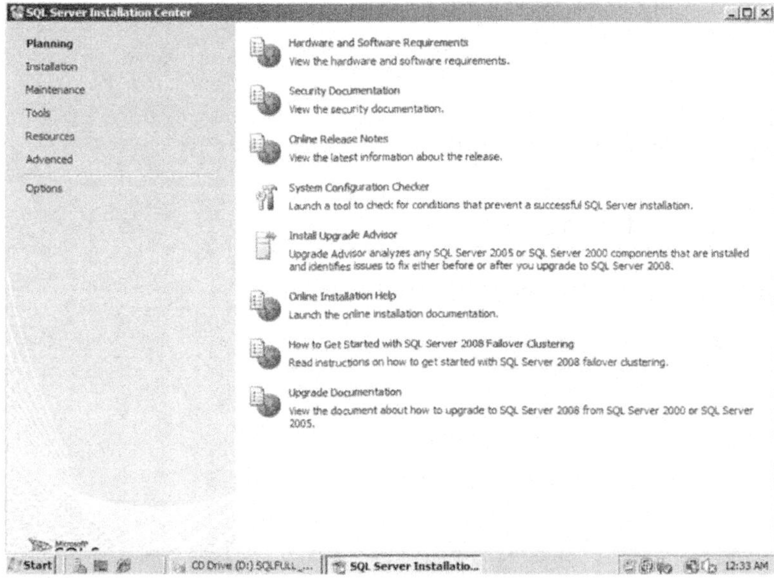

Figure 3-6: *SQL Server Installation Center*

10. Click the first link **New SQL Server stand-alone installation or add features to an existing installation**.

11. Next you see the following screen:

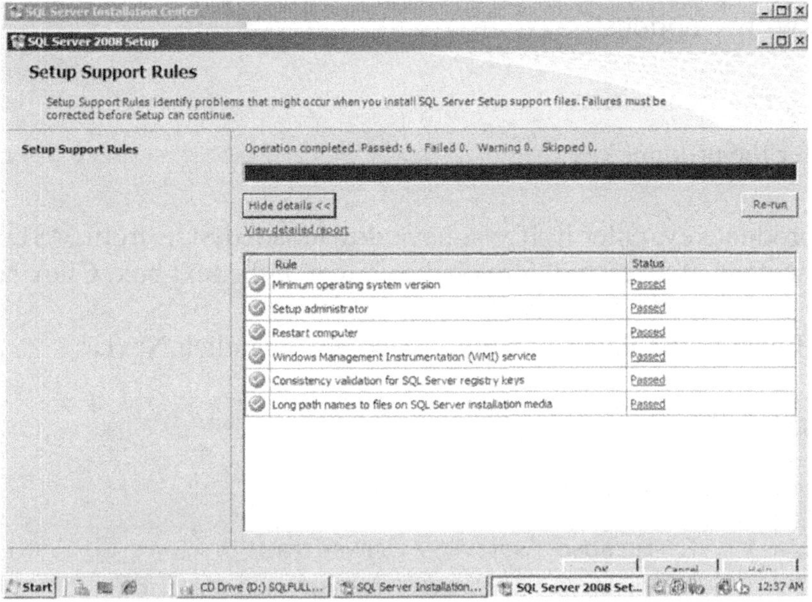

Figure 3-7: *Setup support roles*

12. Click **Ok**.
13. You will see following screen:

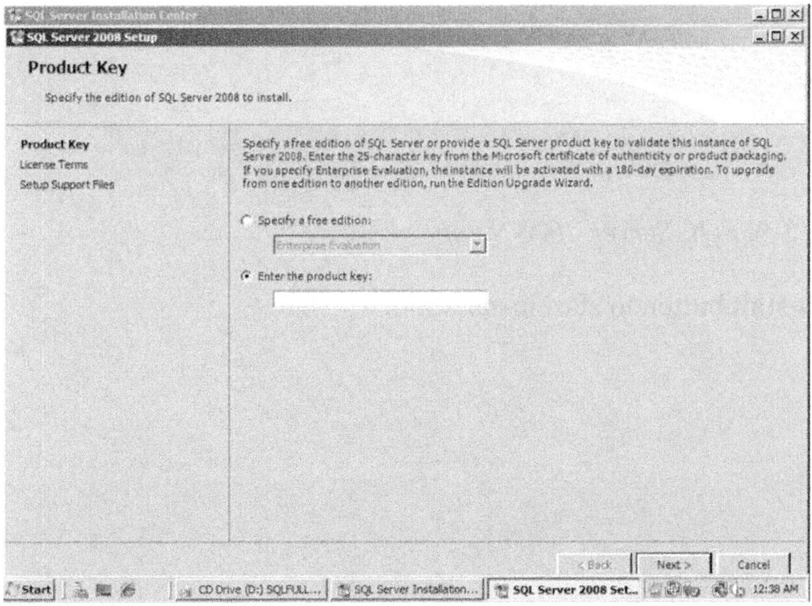

Figure 3-8: *SQLServer product key*

The screen shows two options:

- Specify a free edition
- Enter the product key

If you have a product key, enter it. If you have downloaded SQL from MSDN subscribers download, the key will already be shown in the text box. Click **Next**.

14. Check **I accept the license terms.** Check box and click **Next**.

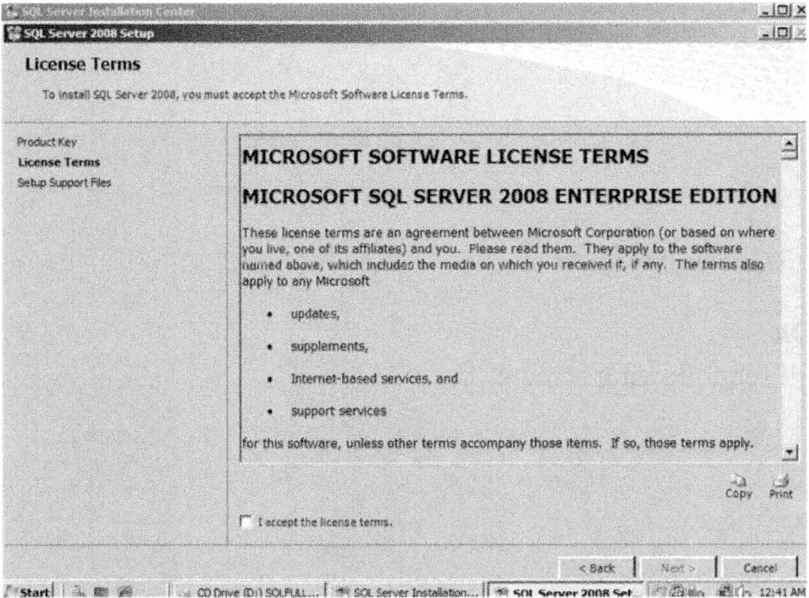

Figure 3-9: *SQL Server 2008 Setup*

15. Click **Install** button to start installation.

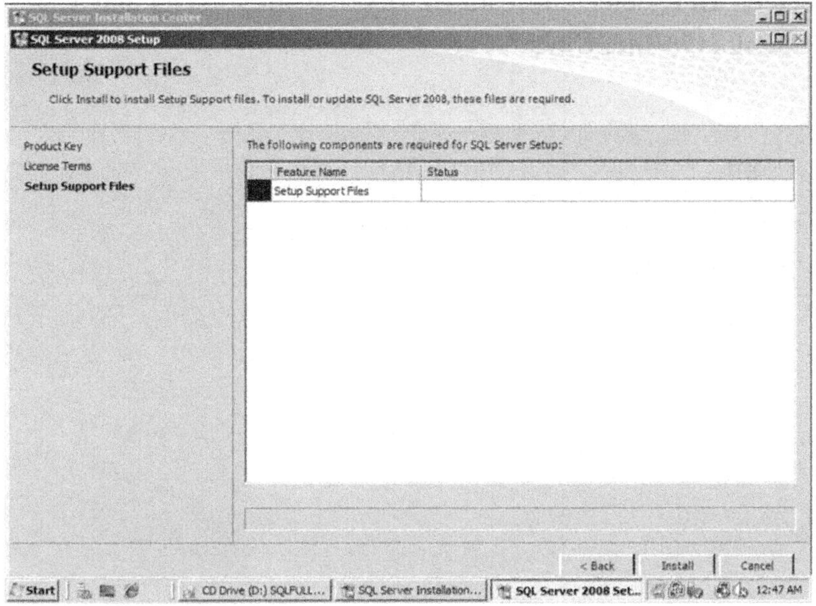

Figure 3-10: *Setup support files*

16. After you see the following screen, click **Next**.

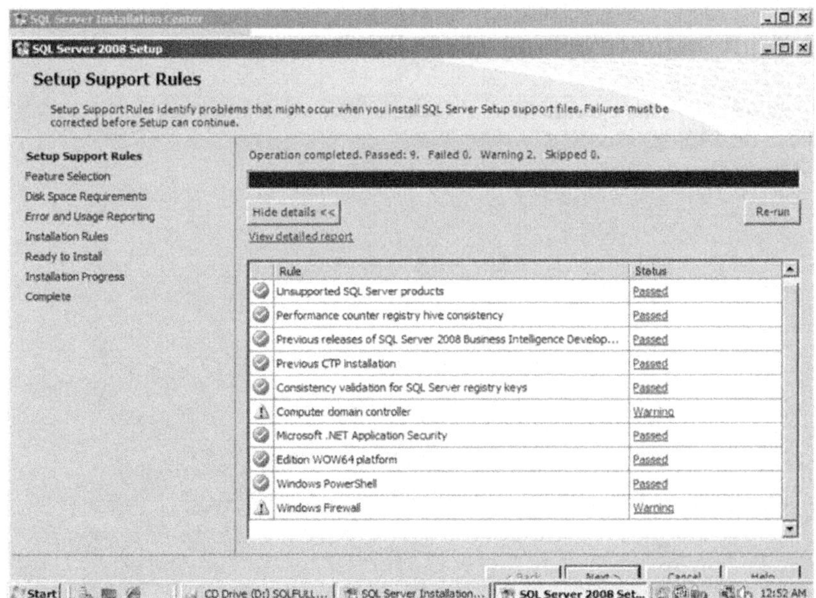

Figure 3-11: *Setup support rules status*

17. Click **Select All** to select all features. If you want you can uncheck **SQL Server Books Online**. Click **Next**.

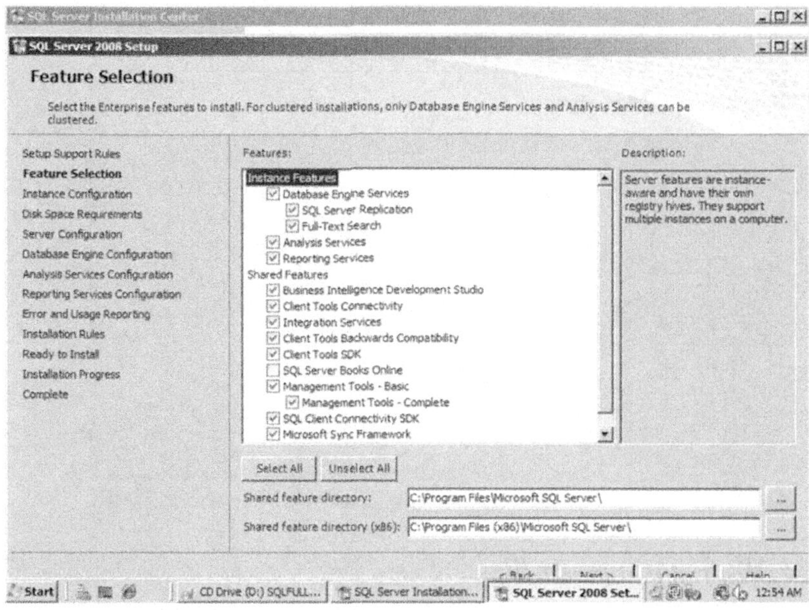

Figure 3-12: *SQL Server 2008 feature selection*

18. Keep **Default instance** option selected and click **Next**.

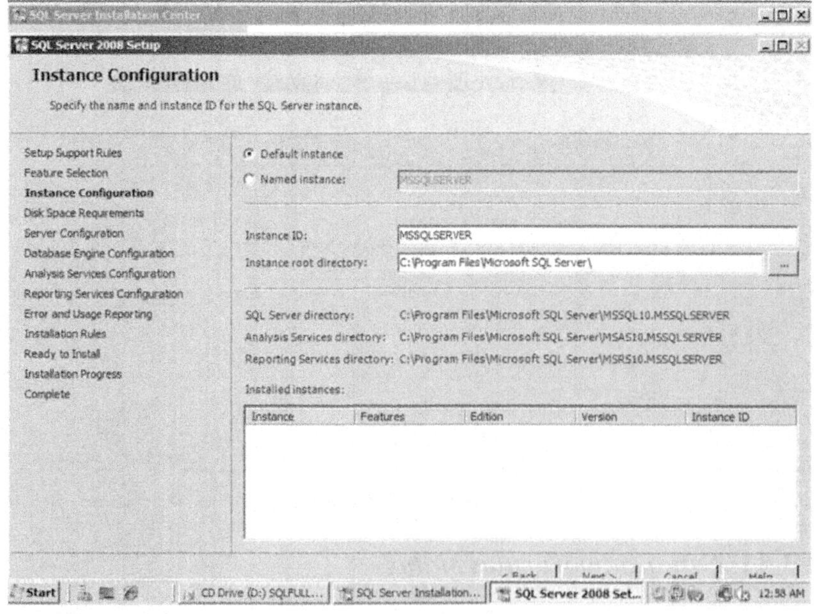

Figure 3-13: *Instance configuration*

19. Click **Next**.

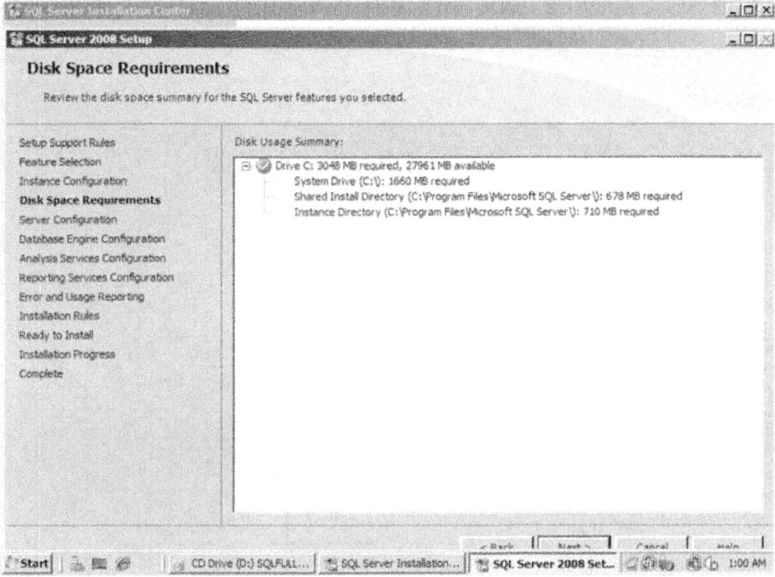

Figure 3-14: *Disk space requirements*

20. Click **Use the same account for all SQL Server services** button and click **Next**. In production environment, you will want to use a separate account for each service.

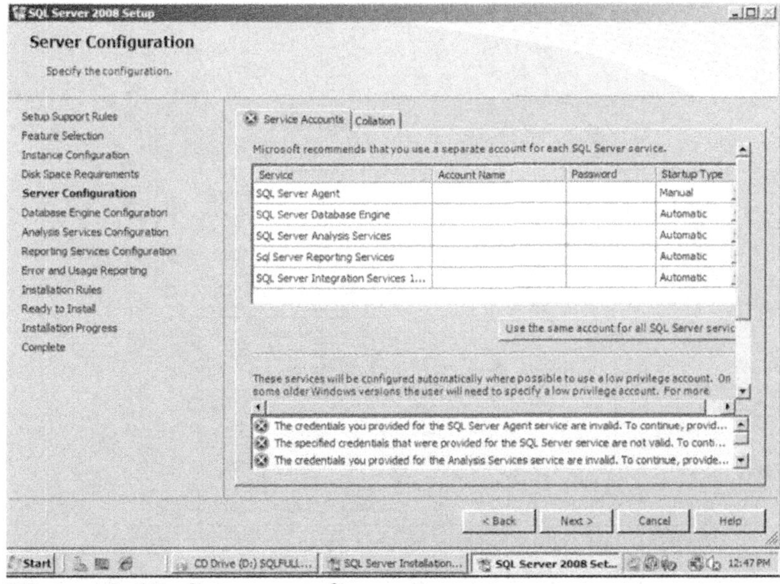

Figure 3-15: *Server configuration*

21. If you remember the account, enter it directly in the **Account Name** or use **Browse** button to search the account name. Enter password and click **Ok**. We created an SQL service account (SQLAccount) earlier specifically for this step.

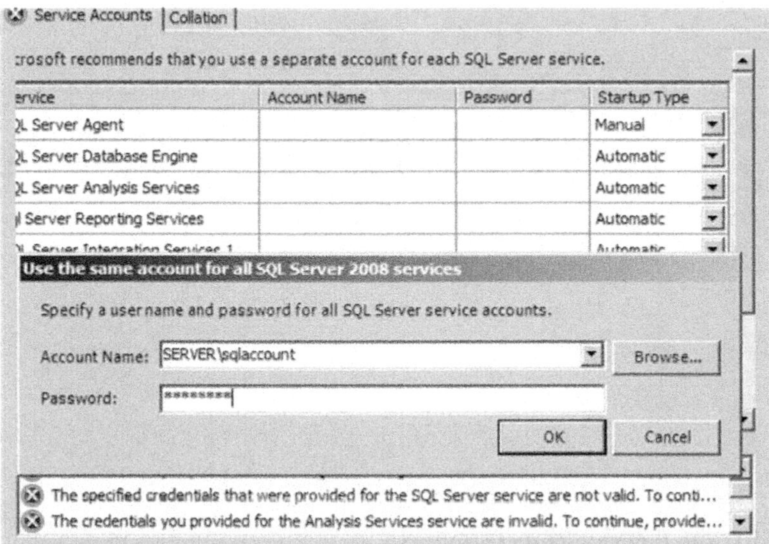

Figure 3-16: *SQL Server 2008 services account*

22. Click **Next**.

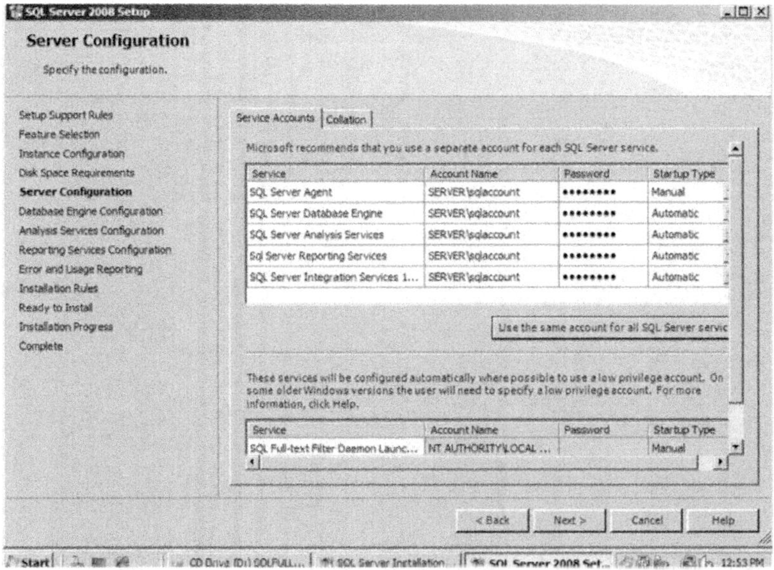

Figure 3-17: *Server configuration – accounts*

23. Click **Add Current User** button. This will add the currently logged in user as SQL Server Administrator. Click **Next**.

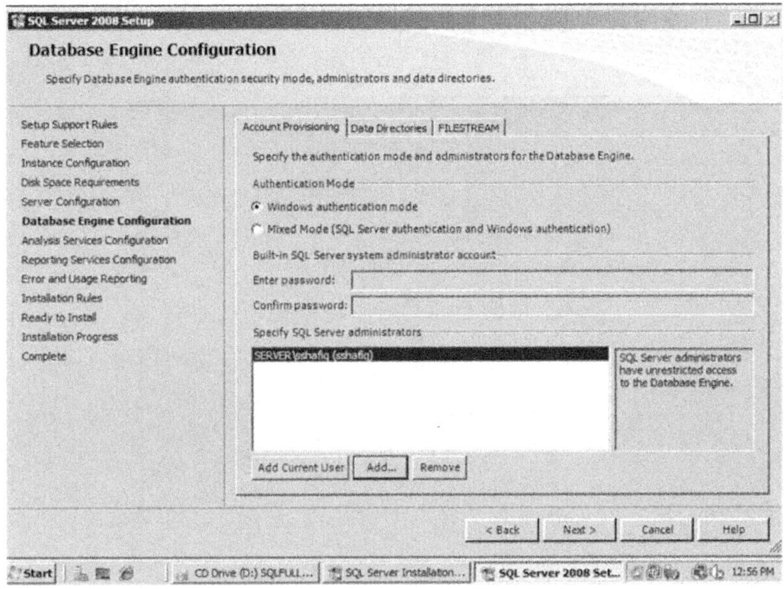

Figure 3-18: *Database engine configuration*

24. Again click **Add Current User** button. This user will have administrative permissions for Analysis Services.

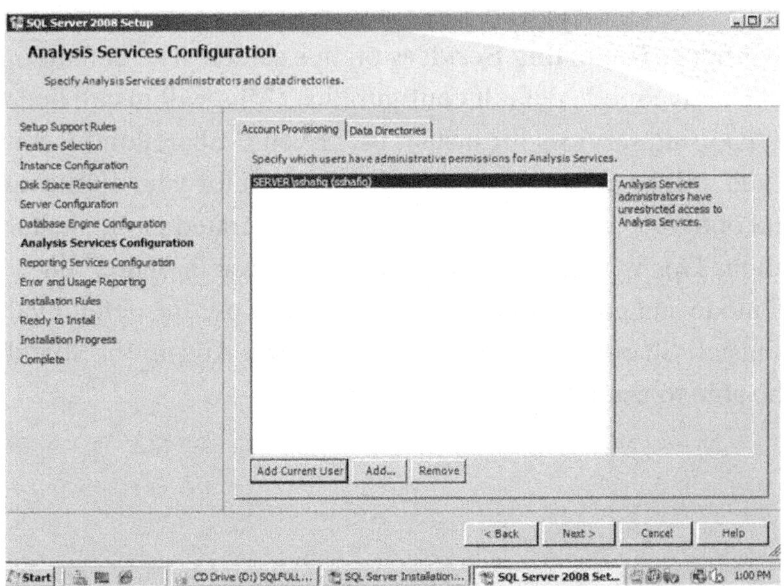

Figure 3-19: *Analysis services configuration*

You can also specify data directories for SQL Server Analysis services. Click on **Data Directories** tab to see the default directories. Click the ellipsis button to browse and select a new directory. You can select directories for data, log, temp and backup.

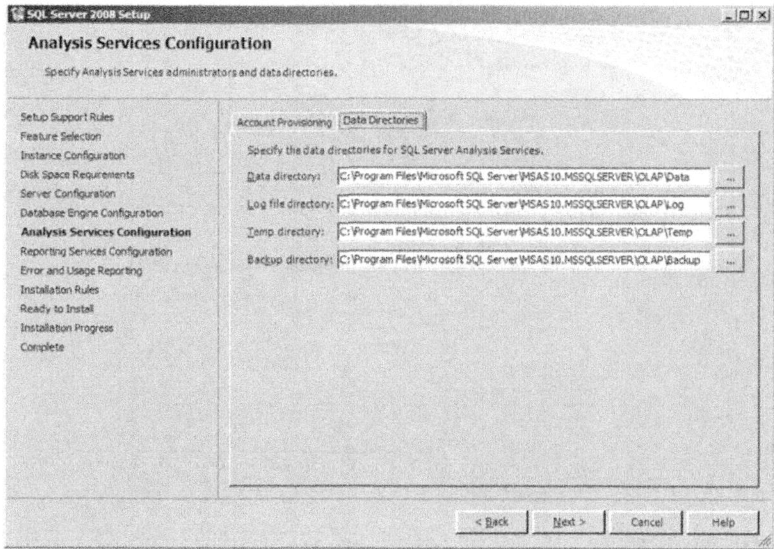

Figure 3-20: *Data directories for SQL Server Analysis Services*

25. Click **Next**.

26. You can configure **Reporting Services** on this screen. The default option is to install the native mode default configuration. This will install and configure reporting services for instant use. Even if SharePoint is not installed, you will be able to use reporting services for your applications. The second option is to install the SharePoint integrated mode default configuration. This will create report server database in SharePoint integrated mode and configure the report server to use the default values. Once you install SharePoint and reporting services Add-In for SharePoint, you will be able to use integrated operations.

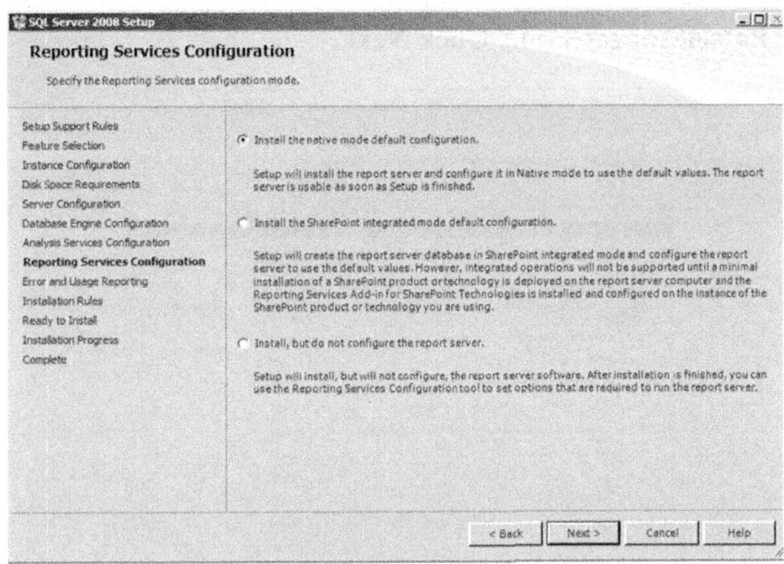

Figure 3-21: *Reporting Services Configuration*

Use the default option **Install the native mode default configuration** and click **Next**.

27. Click **Next**.

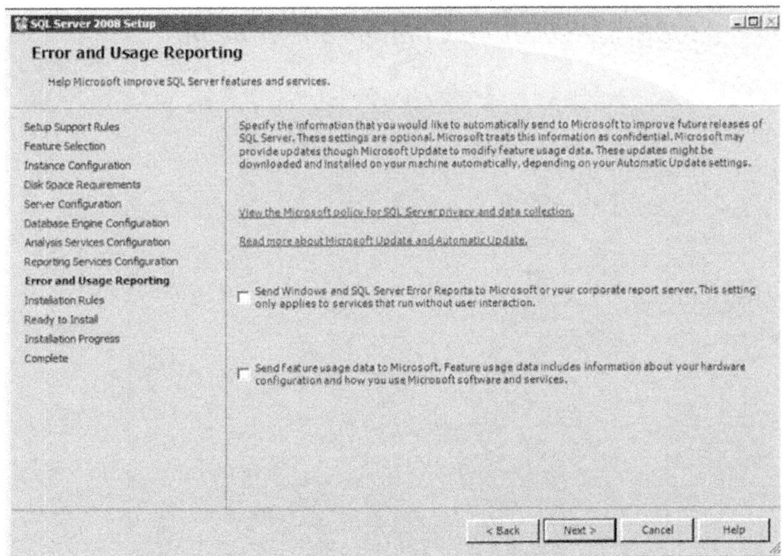

Figure 3-22: *Error and Usage Reporting*

28. Click **Show Details** button. This will show the status of each rule. Status should be **Passed** for each rule. Click **Next**.

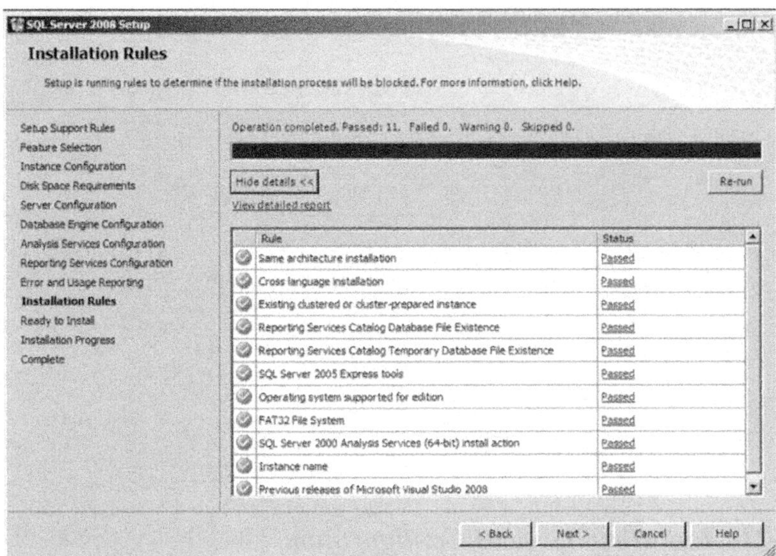

Figure 3-23: *Installation Rules*

29. Configuration is complete. SQL Server is ready to be installed now. Configuration file will be saved at the following path:

C:\Program Files\Microsoft SQL Server\100\Setup Bootstrap\Log\20091228_003646\ConfigurationFile.ini

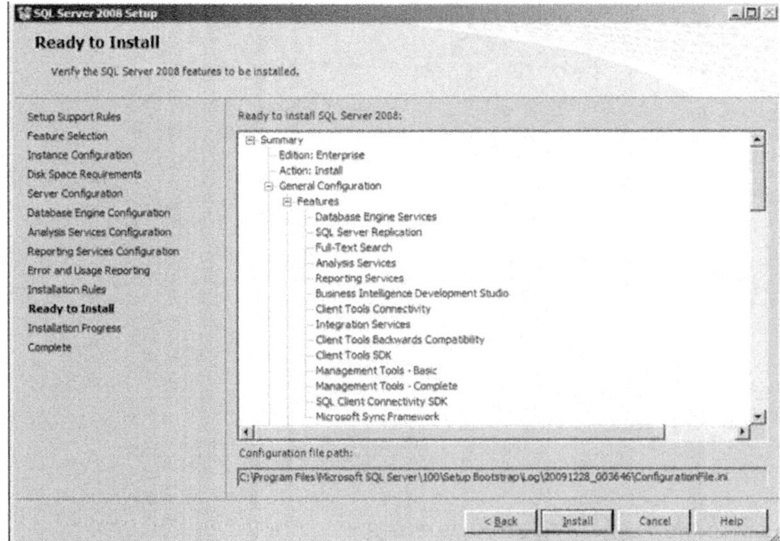

Figure 3-24: *SQL Server is ready to install*

Click **Install**.

Installation will take anywhere from 20 to 50 minutes depending on your computer's speed.

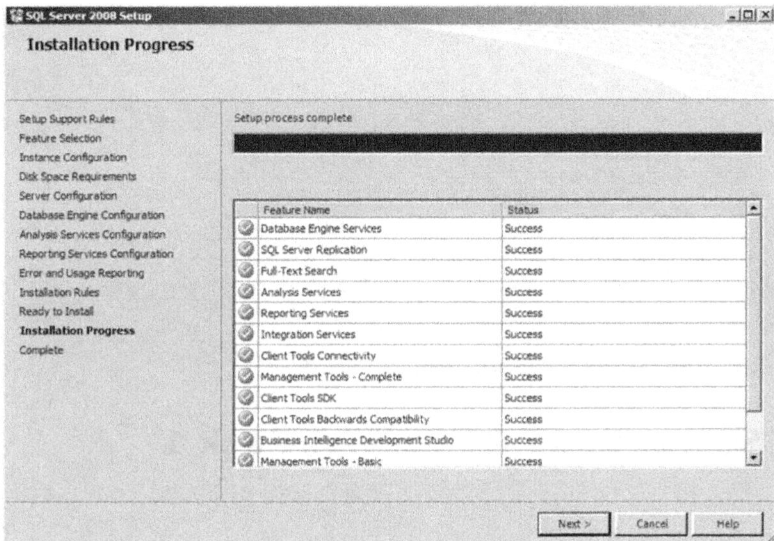

Figure 3-25: *SQL Server installation in progress*

Click **Next**.

30. Following screen will have a link to the summary log file. Supplemental information is also provided on this screen. Click **Close**. Your installation is now complete.

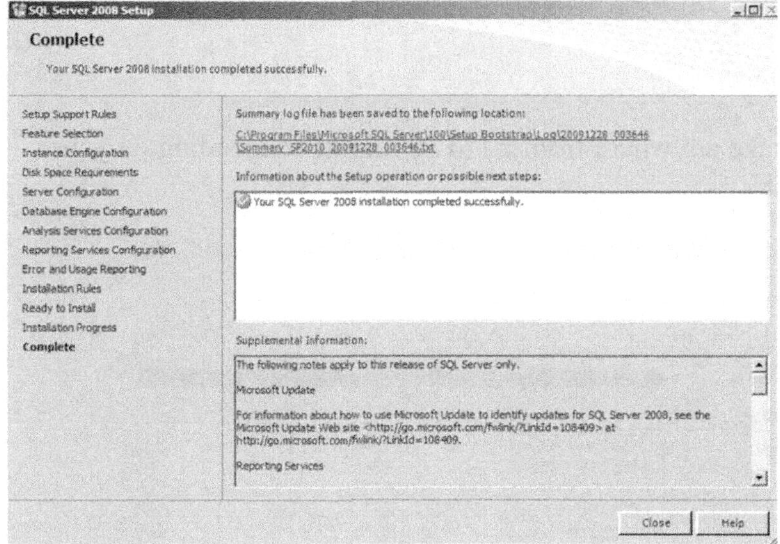

Figure 3-26: *SQL Server installation completed*

Install SQL Server 2008 R2 Service Pack 1

Note: Skip this section if you already have SQL Server 2008 R2 SP 1 or SP 2 installed. SP1 is the minimum requirement for SharePoint 2013.

31. Download Service Pack 1 for SQL Server 2008 R2 from the following link:

    http://www.microsoft.com/en-us/download/details.aspx?id=26727

Download following option:

- SQLServer2008R2SP1-KB2528583-x64-ENU.exe **(309.0 MB)**

We only need service pack 1 for 64-bit systems (fourth option). Click **Download** button to start downloading.

32. Run service pack installation package. It will extract all files to a folder on your system drive. Make sure the logged in user has full privileges to install files. Specially, the user must have write permissions on the system drive otherwise the extraction of files will fail with the message: **Operation was cancelled by the user.**

33. Click **Show Details** to see status of each rule. Click **Next**.

34. Accept the license terms (by checking the **I accept the license terms.** option) and click **Next.**

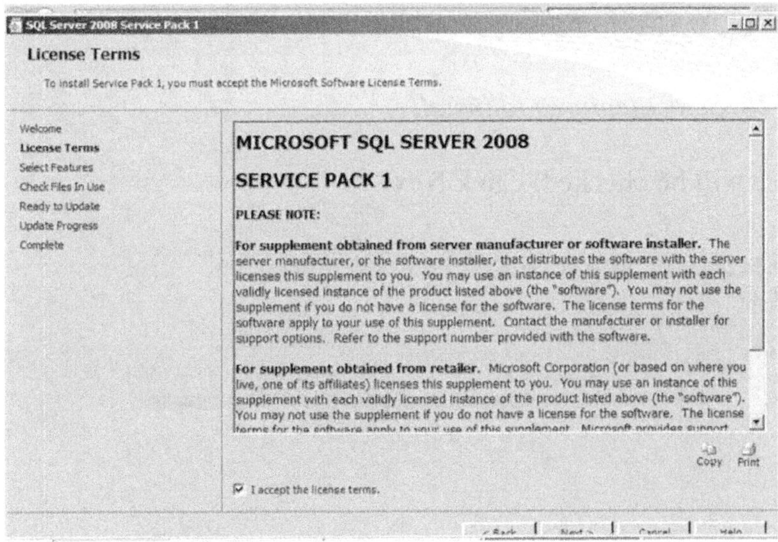

Figure 3-27: *SQL Server 2008 Service Pack 1 License Terms*

35. Click **Next**. Installation will start.

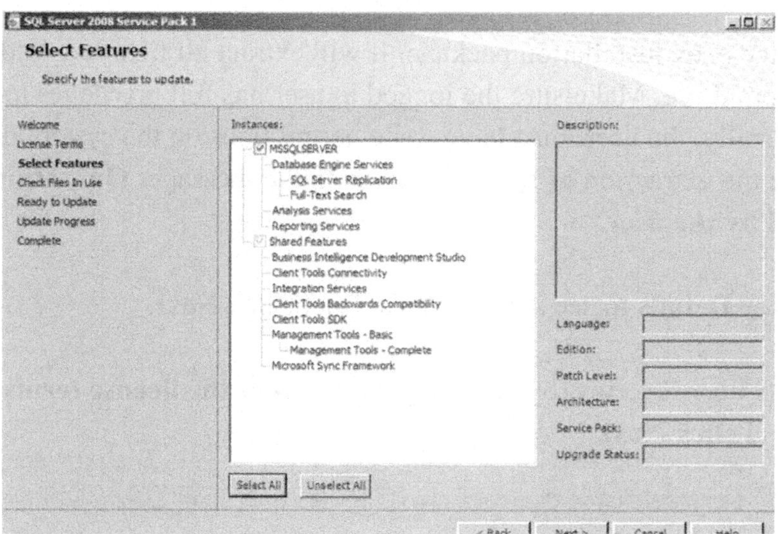

Figure 3-28: *Select features to install*

36. Files in use will be checked. Click **Next**.

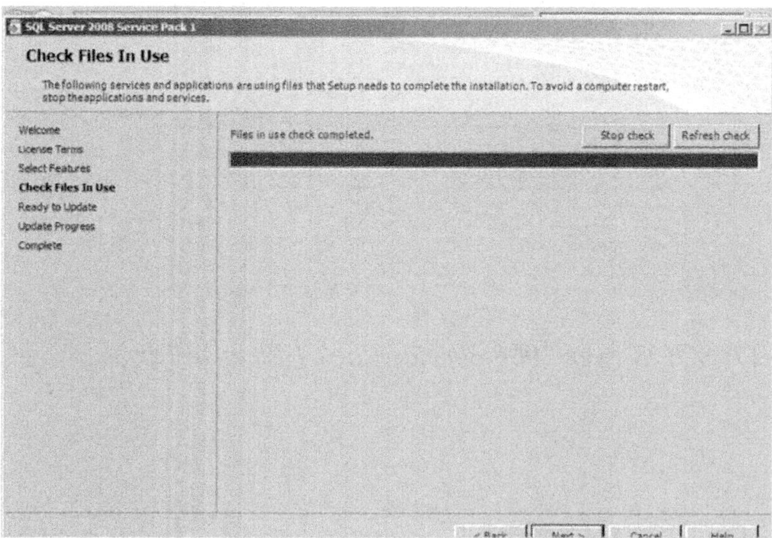

Figure 3-29: *Check files in use*

37. Click **Update** to start the process. Process will take approx. 20 – 40 minutes to complete depending on your computer's speed.

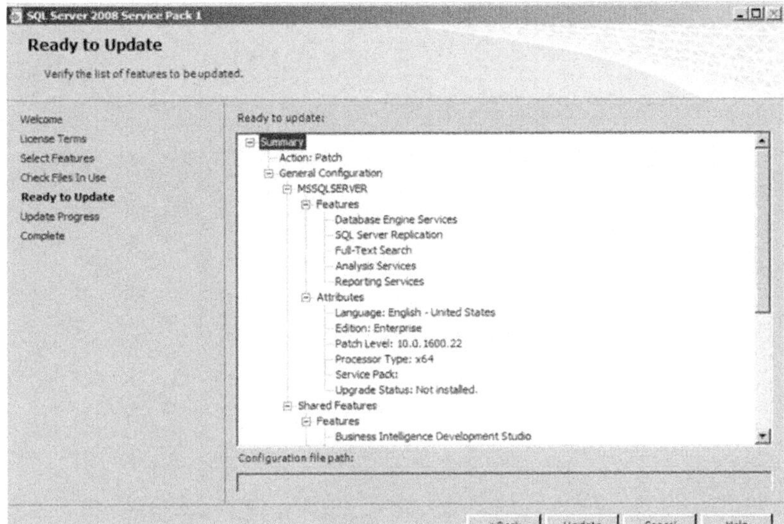

Figure 3-30: *Features to be installed*

38. Click **Next**.

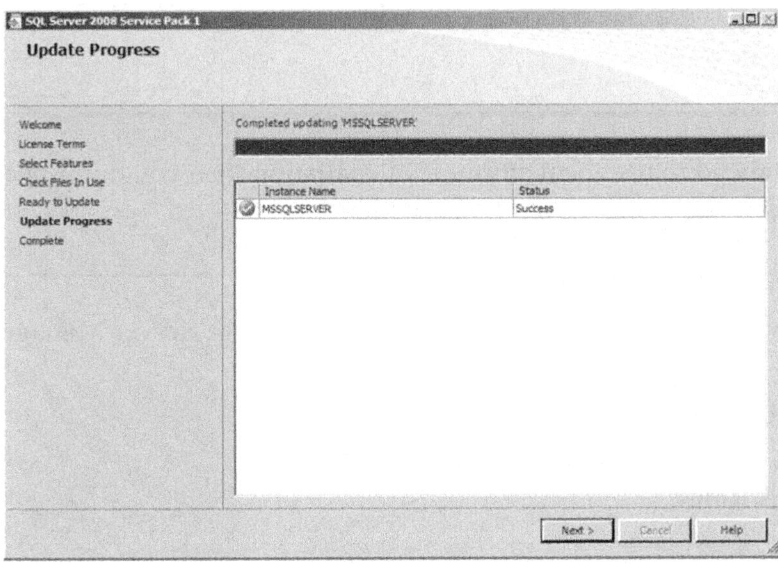

Figure 3-31: *Update progress*

39. Click **Close**.

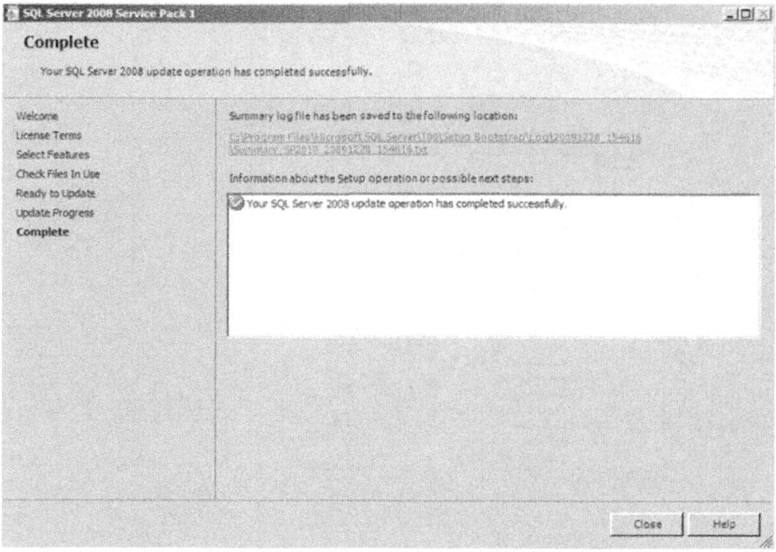

Figure 3-32: *Complete status*

Install Application Server and Web Server (IIS) roles

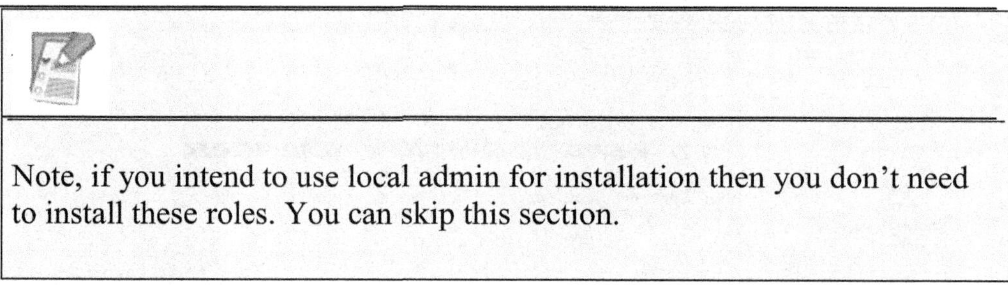

Note, if you intend to use local admin for installation then you don't need to install these roles. You can skip this section.

40. Open server manager (Start > Administrative Tools > Server Manager).

41. Expand **Roles** node.

42. Click **Add Roles**.

43. Click **Next**.

44. Check **Application Server** checkbox.

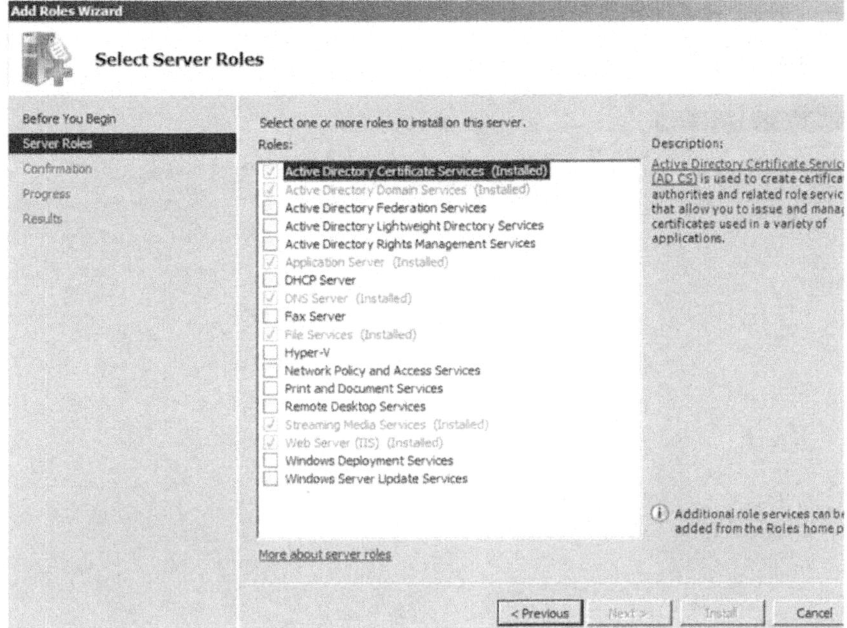

Figure 3-33: *Select Server Roles*

45. Click **Add Required Features** button.
46. Check **Web Server (IIS)** checkbox.
47. Click **Next**.
48. Click **Next**.
49. Check **Web Server (IIS) Support** checkbox.

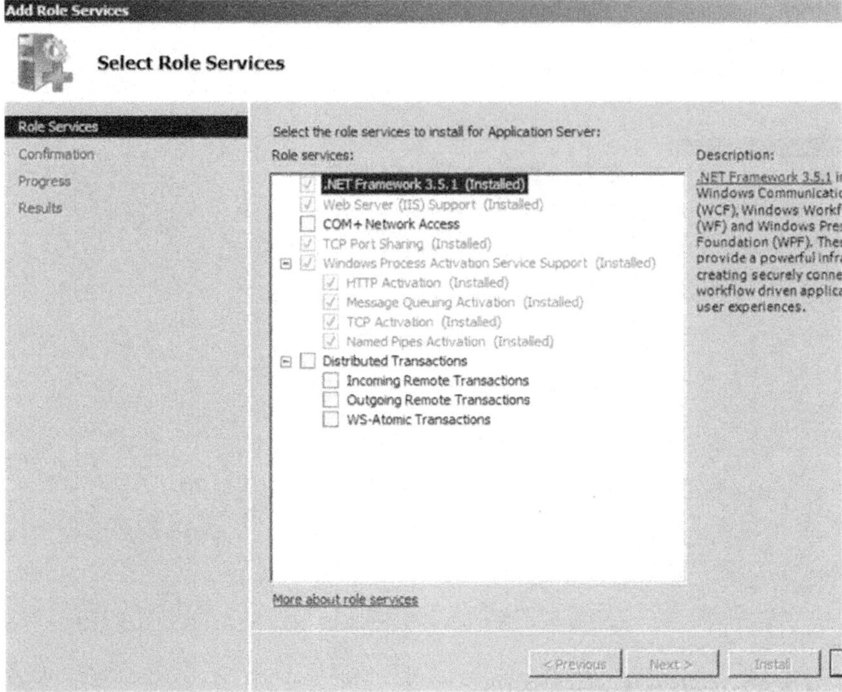

Figure 3-34: *Select role services*

50. **Add Roles Wizard** will pop-up. Click **Add Required Roles Services** button.

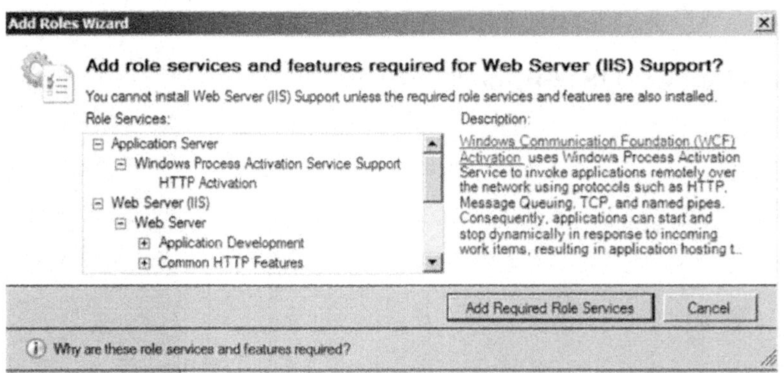

Figure 3-35: *Add roles wizard*

51. Click **Next**.
52. Click **Next**.
53. Click **Next**.

54. Click **Install**.

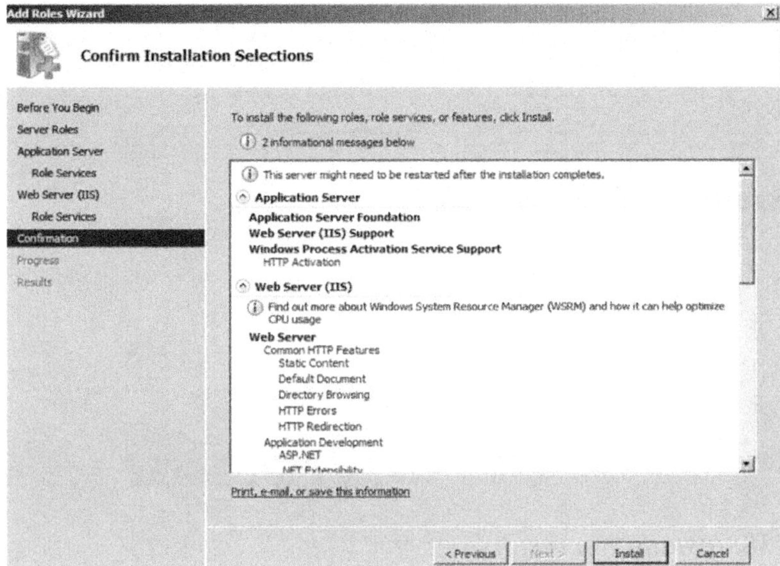

Figure 3-36: *Installation selections*

Installation will take some time so be patient.

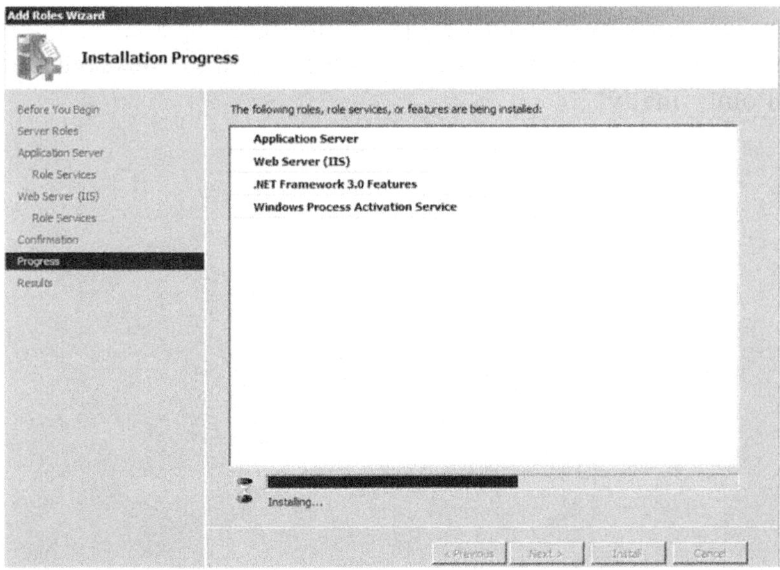

Figure 3-37: *Installation progress*

55. Click **Close** button.

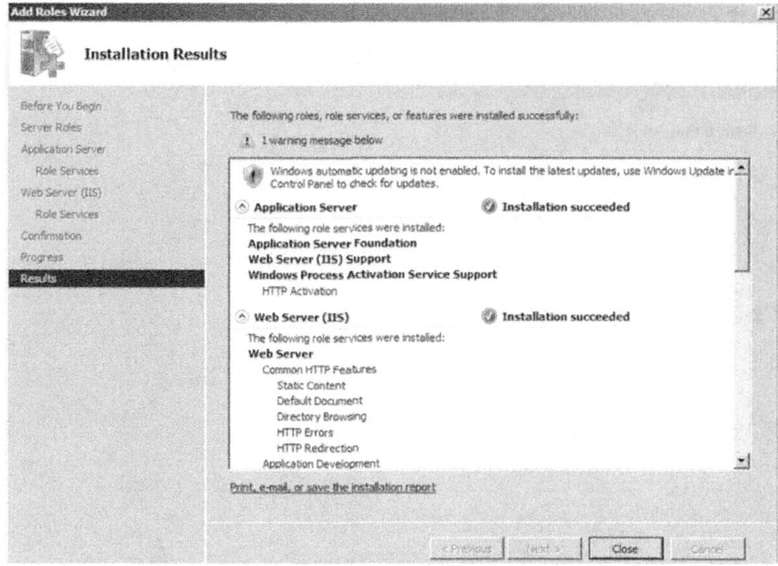

Figure 3-38: *Installation results*

Summary

In this chapter you learned how to install the pre-requisites, especially, .NET framework and SQL server and SQL Server service pack. In the next chapter, you will install SharePoint, finally!

Chapter 4

Install SharePoint 2013

In this chapter:

- Download and Install SharePoint
- Configure SharePoint
- Create Site Collection

Download and Install SharePoint 2013

1. Download SharePoint 2013. You can download it from MSDN if you are a subscriber.

2. Run prerequisiteinstaller.exe to install prerequisites.

3. Accept agreement and click **Next**. Make sure you are connected to the Internet. Prerequisites will be downloaded and installed on your machine. Basic prerequisites will be installed and you will be asked to restart the machine. Click **Finish** to restart the machine.

4. After the restart, installer will continue to install the prerequisites. You will see **Installation Complete** screen when installation has completed. Click **Finish**.

Restart your machine and then run **setup.exe**. It will ask for product key. If you did not download product key when you downloaded SharePoint 2013, then log back into MSDN and download the key. Once you have the key, enter it.

5. Accept the terms and Click **Continue**.

6. Choose the default locations and click **Install Now**. If you want to change the installation location, you can do so now. Software starts installing.

Configuring SharePoint

7. After the installation, run configuration wizard.

8. Click **Next**. It will prompt you to restart IIS and other services. Click **Yes**.

9. Select second option **Create a new server farm** and click **Next**.

10. Enter Database Server. Usually SQL server name is same as machine name. If your machine name is SP2013 then your SQL Server name will also be SP2013. Enter database access account details (username and password). Click **Next**.

11. Enter passphrase in both boxes and click **Next**. Passphrase can be anything. I prefer something that can be easily remembered. Whatever you decide, write it down. You may need it in future.

12. Check port number checkbox if you want to change the port otherwise leave it unchecked. Keep "NTLM" selected and click **Next**.

13. Click **Next**. Configuration will start and will take some time to complete. Click **Finish**. A browser window will open and prompt you to enter user credentials. Enter admin credentials and click **OK**. Browser will load SharePoint Central Admin site.

14. Click **Ok**. Click **Start the Wizard** button. You can also do configuration manually. To do manual configuration click **Cancel**. For this exercise, we will use the wizard.

15. In this step we are going to use the services account (SPService) that we created in the beginning when setting up Active Directory. Select second option **Create new managed account** and enter services account information (username/password) in the username and password fields. Leave the default services selected. If you want to add or remove any service, check or uncheck it on this page. Click **Next**. You will get an error if you entered incorrect username.

16. After the basic configuration, you see following **Create Site Collection** screen.

Figure 4-1: *Create site collection*

Enter a title and description. Select /sites/ from the drop down in the Web Site Address section and enter a name for your site. This will create a site collection under **/sites**. Select **2013** in the **Select experience version**. This will base your site collection on the 2013 template. You can also select **2010** if you want to base your site collection on the **2010** template. If you want to create a site collection on the root, select "/" from the drop down. Select **Team Site** under the **Collaboration** tab and click **OK**.

17. You will see following message and screen after your site has been created:

Figure 4-2: *Configuration completed*

18. Click on the site URL to open the site. Enter userid and password. Your site will look like this:

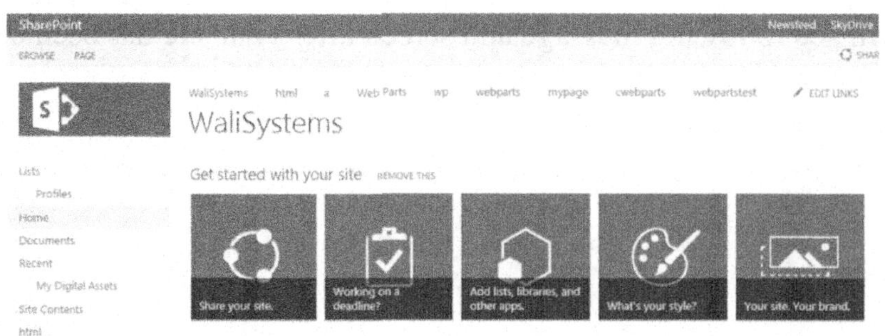

Figure 4-3: *SharePoint site*

Congratulations, you have just created your first SharePoint 2013 site.

Renaming SharePoint Server

This is a bonus section. It is a common scenario and considered to be the most complicated one. You change the name of the machine where SharePoint 2013 is installed and you break everything. Yes, you read it right. I am talking about the machine name change. So what are the steps involved in changing the machine name that has SharePoint 2013 installed?

1. First thing that I would suggest is that you use a local account for all the steps mentioned below. You can also use a domain account with administrator privileges but using a local account is preferable. If your machine is connected to the domain, that increases the headaches. I would recommend, disconnect it from the domain and then use local admin account to configure everything. Once everything is setup correctly and running, you can join the domain again. If you have already changed the name of the machine and are now facing issues, just rename it back to what it was before. It may not allow you to change the name back if you are logged in as local admin and still connected to the domain. So login with domain admin account, disconnect the machine from the domain and then log in again with local admin account. Now change the machine name back to what it was. Complete your configuration and then join the domain again but before you change the machine name there are a few other steps that you have to take.

2. So, are you ready? OK. If you have not changed the machine name yet, that's great. Make sure "Builtin\Administrators" group is added to the SQL Security this is because after you change the machine name you will have trouble logging in so it's important that local admins have permissions in the SQL server.

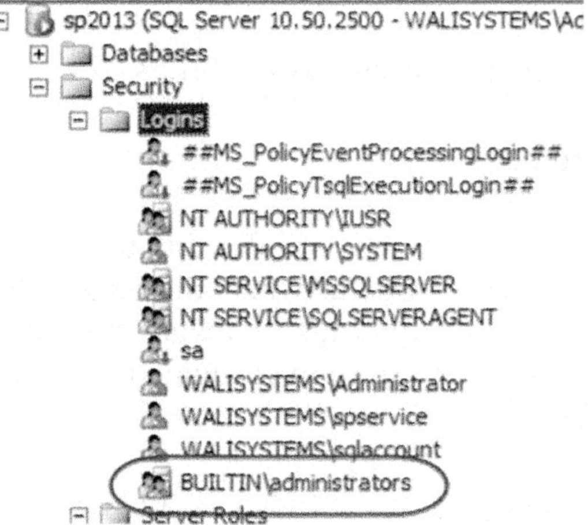

Figure 4-4: *Add Builtin\Administrators to SQL Logins*

Next, make sure that “Builtin\Administrators” are added to the “sysadmin” role.

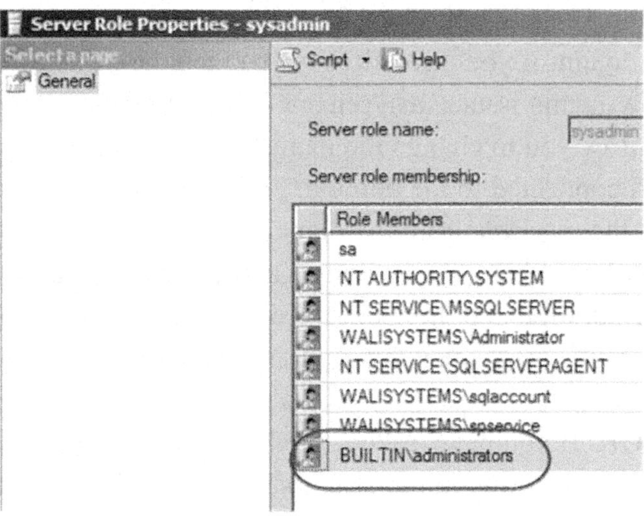

Figure 4-5: *SQL Server Roles*

Note: If you are not using local account to make changes, then make sure the domain account you are using is added in SQL logins and also in sysadmin role.

3. Now open “SharePoint 2013 Management Shell”. Make sure you use elevated privileges to run the shell otherwise the command that you are going to run below will fail.

4. Run following command in “SharePoint 2013 Management Shell”:

rename-spserver –identity "oldname" –name "newname"

"oldname" is the original machine name that your computer had and "newname" is the new name. Remember, you have changed the machine name, now goal is to replace old machine name with new name in SharePoint databases. The above command will make necessary changes in the databases.

You may get following errors when trying to run the above command:

Cannot access the local farm.

OR

Rename-SPServer : The farm is unavailable

If you get this error, run following command in management shell:

get-spshelladmin

See if the account you are using is listed in the results returned by this command. If your account is not listed, run following command to add it to the list:

add-spshelladmin "accountname"

Replace "accountname" with your actual account name. Now you should be able to run the rename command as shown above. Note, you may get following error after running the command:

Rename-SPServer : Object reference not set to an instance of an object.

This may not necessarily mean that the action was unsuccessful. Even though command throws this error but server name is changed successfully. To verify, you can log in to central administration site and check there but there are a few more steps to follow before you can access central admin site.

Note: You may get following error on trying to rename the server name:

The local farm is not accessible. Cmdlets with FeatureDependencyId are not registered.

Reason: You have already changed the machine name and now when you run the rename command, it does not find the local farm because name has changed. So change the machine name back to what it was originally and then run the rename command again.

5. Now change the machine name. After you have changed the name, open SQL management studio and try to connect to the database. If you are unable to connect, don't worry. Wait for a few minutes. Machine name change takes some time to take effect. If you have succeeded in connecting to the database, proceed to the next step.

If you are confused about using local and domain accounts and are having trouble in connecting to the database and you have already changed the machine name, revert back the (machine name) change. Connect to the SQL Server and create an SA account. Make sure you give SA account permissions on the "master" database. Now change the machine name and if you have trouble connecting to SQL using "Windows Authentication", use "SQL Authentication".

6. You were not supposed to do anything in the previous step apart from ensuring that you could access the SQL server. Run "SharePoint 2013 Products Configuration Wizard". If wizard fails, try PowerShell command. Open "SharePoint 2013 Management Shell" and run following command:

psconfig.exe -cmd upgrade -inplace b2b -wait –force

Note: Run management shell with administrator privileges otherwise the upgrade may fail. To run it with admin privileges, right-click the "SharePoint 2013 Management Shell" command and select "Run as administrator".

If it still fails, open the log file and see what's causing the trouble. Upon failure, configuration wizard will create a log file and show you the exact path on screen. Just click that link to open the file. Just in case, if you are wondering, the log file location is as following:

***System Drive**:\Program Files\Common Files\Microsoft Shared\Web Server Extensions\15\LOGS*

You will see something like this at the bottom of the file:

Last message from task upgrade is An exception of type Microsoft.SharePoint.Administration.SPUpdatedConcurrencyException was thrown. Additional exception information: An update conflict has occurred, and you must re-try this action.

Usually this error occurs when you don't have correct permissions. Running wizard with admin privileges helps you avoid this problem but you never know, if it still occurs then follow the tips given below.

Tip #1: Tip to resolve concurrency exception

Visit following KB article. It has the solution to this problem.

http://support.microsoft.com/kb/944267

KB article gives following cause and resolution:

Cause:

This issue occurs because a lock exists on the SPFarm object. Therefore, the SharePoint Products and Technologies Configuration Wizard cannot update the databases.

Resolution:

To resolve this issue, clear the Windows SharePoint Services configuration cache. If you need exact steps to clear the cache, visit following KB article:

http://support.microsoft.com/kb/939308

This KB article is for SharePoint 2007 but with minor changes you can make it work for 2013 as well. For example, there is no "Documents and Settings" folder in Windows Server 2008. It's the "Users" folder on the system drive. In Windows Server 2008, the configuration cache is in the following location:

***System Drive**:\ProgramData\Microsoft\SharePoint\Config\GUID*

Don't delete the cache.ini file. Just clear it. Select all data in the file and delete it. Then run the wizard again. You will see success message upon successful completion.

7. If you try to access SharePoint site in browser now, using old URL, you will get page not found. If your site URL contains the machine name, replace that part with the new machine name and try again. Now you may see another error "An unexpected error has occurred". Now, you should be happy that you have reached this far. At least you see a familiar SharePoint error now. Next part is to fix the sites.

8. Now open central admin site. Please note that your old URL will not work. To open central admin, just replace the part before the port number with **localhost**. The site should open now. For example, if central admin URL was **http://serverA:13214**. Replace **serverA** with **localhost** so the URL will be **http://localhost:13214**.

9. Change Alternate Access Mappings. In central administration site, in **System Settings**, under **Farm Management**, click **Configure alternate access mappings**.

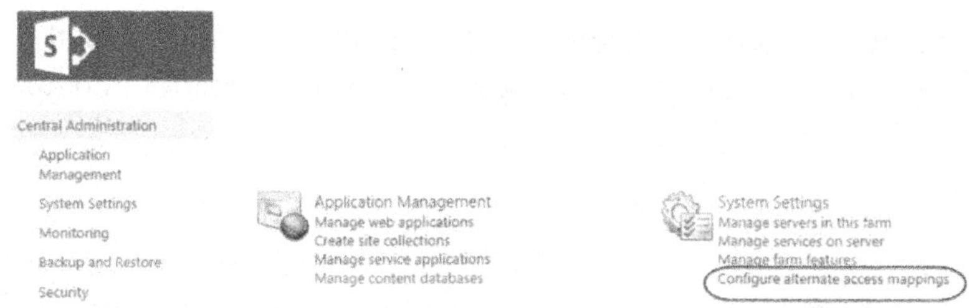

Figure 4-6: *Alternate Access Mappings*

10. Click **Edit Public URLs**. If you don't see any URLs, change **Alternate Access Mapping Collection** from the drop down. Now change the URLs. Replace old server name with the new one. After this change, you will be able to access central admin site with the server name in the URL instead of using localhost.

11. Enable detailed error messages in web.config. This is important because if SharePoint site loading fails, you should be able to see a detailed error message. Of course, you can also check error logs for details but it is convenient to see error messages instantly when the problem occurs. To enable error messages, follow the steps below. By default, you see following message when something goes wrong:

Server Error in "/" application.

This is not a helpful message. To enable detailed error message, do the following:

1. In your application's web.config, locate **<CustomErrors ...** tag and set the mode to **Off**. The tag will look like this after the change:

<customErrors mode="Off" />

2. Locate **<SafeMode MaxControls="200" CallStack=...** tag and set the value of **CallStack** attribute to **true**. The tag will look like this after the change:

<SafeMode MaxControls="200" CallStack="true" DirectFileDependencies="10" TotalFileDependencies="50" AllowPageLevelTrace="false">

If you still don't see detailed error then go to the following path:

System Drive: *\Program Files\Common Files\Microsoft Shared\Web Server Extensions\15\TEMPLATE\LAYOUTS*

Open **web.config** and set **CustomErrors** to **Off** (just like you did above).

Setting customErrors mode to Off is enough to show detailed messages but in case if it still does not work for you, do the following as well.

In same **web.config**, set **CallStack** attribute of **<SafeMode>** tag to **true**. If you don't find that tag, add the following under **<Configuration>** tag:

<SharePoint>
<SafeMode MaxControls="200" CallStack="true" DirectFileDependencies="10" TotalFileDependencies="50" AllowPageLevelTrace="false">
<PageParserPaths>
</PageParserPaths>

</SafeMode>
</SharePoint>

Now you should be able to see detailed error messages.

12. Now the last step! Existing sites may not load because of a problem with user profile service application. New sites that you create work perfectly fine. Obviously, one solution is to recreate your sites. That sounds easier but it's not practical especially if you have many sites with customizations, data, users, etc. If you have few sites that were created without any heavy customizations, go ahead and recreate the sites otherwise continue reading and see how you can resolve this issue. You see following (detailed) error when you try to load an existing site in browser.

Server Error in '/' Application.
Object reference not set to an instance of an object.

You get this error because the account name has also changed with the machine name change. For example, if your account name prior to machine name change was "machine1\admin", now it would be "machine2\admin" (assuming new machine name is "machine2"). You have to give permissions to the new account name in User Profile Service Application. How do you do that? Continue reading!!!

13. In central administration site, click **Application Management**.

14. Click **Manage service applications** (under **Service Applications**).

15. Click **User Profile Service Application**. Do not click link, it will take you to a different page. Click somewhere on the right side of the row to select/highlight the row then click **Permissions** button in the ribbon. Also, note that this service application may be named differently in your environment. This is the default name by the way. If you have not changed it, then this is the one you need to modify. Also note that there can be more than one User Profile Service Application. You have to modify the one that is associated with your web application.

To find out which User Profile Service Application is being used by your web application, go to **Application Management** and **click Manage web applications**. Select your application. Click **Service Connections** button in the ribbon. Here you can see which service is associated with your web application. Scroll down to locate the **User Profile Service Application**. There can be multiple user profile service applications listed here. In that case, you will modify each of them or remove the one that are not needed. If you want to remove a service application, from the top drop down in the dialog box, choose "[custom]". That will enable the check boxes. Then uncheck the one you want to remove and click **OK**.

16. Now you know the service name. Go back to **Manage Service Applications** page and select the service and click **Permissions** button in the ribbon.

17. Check if the account you are currently logged in with is showing up in the users' list. You would also get User Profile error if this list shows a user that does not exist anymore.

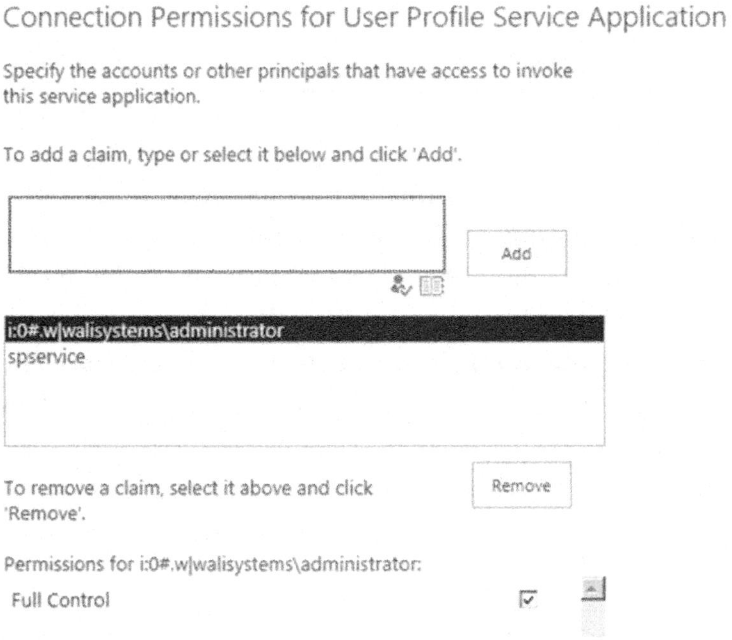

Connection Permissions for User Profile Service Application

Specify the accounts or other principals that have access to invoke this service application.

To add a claim, type or select it below and click 'Add'.

Add

i:0#.w|walisystems\administrator
spservice

To remove a claim, select it above and click 'Remove'.

Remove

Permissions for i:0#.w|walisystems\administrator:

Full Control

Figure 4-7: *Users in User Profile Service Application Permissions*

There are two options to fix the problem. One is to add the user that existed before (and is showing up in the list) in the Active Directory. Second option is to remove the user from this list. Depending on your situation, take the appropriate step. If the user account that is listed here is not in use anymore then the best option is to remove him from the list. Chances are that you are confronted with the same common problem that everyone faces after the machine name change. I briefly touched this issue above. The problem is that account name changes with the machine name change. If your account was “machine1\admin” before the machine name change, it became “machine2\admin” after the name change and the user list we are discussing contains the old name that does not exist. User account change is reflected in the Active Directory. If you go to Active Directory, you will see the new account name (“machine2/admin” or simply “admin”) listed there. Now the problem is you are still using that account and you may be logged in with that account even when making these changes. Common sense tells that by adding that new username, problem will be resolved so you try to add the new username in the list and then you are confronted with another problem. You get following error:

The specified user or domain group was not found

So before you attempt to add the user in the list, let me show you how to remove the user if you opt to go with the first option (of removing non-existing user). It’s simple! Select the user and click “Remove” button. Just in case, if you get an error trying to remove the user through the GUI, use the approach that I am going to show you below for modifying the user.

18. Problem occurs because User Profile Service Application does not find the account name that is listed in the “Permissions”. As mentioned in the previous step, we can fix the problem by renaming the user account name but how do you do that, that is the question! Answer is to change the user name directly in the database which is something I would not recommend but if you are desperate to make things work then go ahead and follow the steps given below. Before we move forward, let me tell you that there is another option that will fix the problem without modifying the database directly and that is to create a new user profile service application and associate it with the web application that throws error. This is simple but may not be an option for you if you were using user profile service heavily before changing the machine name. I will list the steps any way just in case you want to take this route.

19. This step shows how to create new user profile service and associate it with your web application. In Central Administration > Application Management, click Manage **service applications** (under **Service Applications**).

20. Click **New** button in the ribbon and select **User Profile Service Application**.

21. Give a name to the new service. You can use an existing application pool or create a new one for this service. You can keep all other default values OR you can copy the options from the existing User Profile Service Application. Click **OK** to save the changes.

22. Now associate the new service application with your web application. Go to **Manage web applications**. Select the web application you want to change and click **Service Connections** in the ribbon. You will see new service application listed there. Check the new one and uncheck the one that has problem and click **OK**. Now open your SharePoint site and voila! It loads!

23. Now let's discuss the other option. The second option was to change the user account in the database. This will allow you to use your existing User Profile Service Application. Open **SQL Server Management Studio**. Open **Query Editor**. Run following SQL in the query editor:

SELECT [Id] ,[Name] ,[Properties]
*FROM [**SharePoint_Config**].[dbo].[Objects]*
where [name] = 'User Profile Service Application'

Change the database name before running the query. The name may be different on your server but it will have word "_Config" in it. For example, I have different SharePoint instances on my server and the configuration database names on my server are as following:

- SharePoint_Config
- SharePoint2013_Config
- SharePoint2013_2_Config

So, choose the one that is relevant to your SharePoint instance. You are querying the **Objects** table in the configuration database. **Where** clause allows you to see only **User Profile Service Application** entries. There may be more than one entry in the results. In first row, click the value in the **Properties** column. Right-click and select **Copy**. **Paste** it in a notepad. Search for attribute **m_SerializedAcl**. If you find it, look at the user list. It should be same as the one you saw in GUI. If this is not the same list, continue with other entries in the table. There will be entries where you will not even find this attribute. There is another attribute called **m_SerializedAdministrationAcl**. That is different. We are only concerned with **m_SerializedAcl**. Once you have found the correct entry, note down the Id from the **Id** column. We will use it in the update query next.

This is how the tag containing **m_SerializedAcl** attribute looks like:

<sFld type="String" name="m_SerializedAcl">

This tag contains all the user accounts. Locate the one that is causing the trouble and modify it. For example, if the account name (causing the problem) was "machine1/admin". Change it to "machine2/admin". Remember "machine2" represents your new machine name. Now copy the whole text in the notepad. Enter following update query in the query editor:

update ***[SharePoint_Config]****.[dbo].[Objects] set Properties=****'Paste updated TEXT copied from Notepad here'*** *where Id=****'772ACCBD-3DF1-430F-82A6-5DB47598B14F'***

You are updating the **Objects** table. Replace database name with the correct database name (relevant to your instance) in the query before running it. Also change **Id** in the **where** clause. This is the same **Id** that you noted down when you found the entry (to be modified) in the table. We will update the value in the **Properties** field with the modified text copied from the Notepad. Run the query and you are all set. Open the SharePoint site and it should load fine now.

That's it! Changing machine name that has SharePoint installed on it is a cumbersome process, especially if you use machine name in the FQDN.

Summary

In this chapter downloading and installing SharePoint 2013 was discussed. You also learned how to create a site collection. Site collection is a top level site in SharePoint and can contain subsites. As the name suggests, it is a collection of sites. Each site collection can have its own URL. You also learned how to change machine name that has SharePoint installed on it.

Chapter 5

Web Enabled Electronic Forms

In this chapter:

- Create Web-Enabled Electronic Forms
- Layout InfoPath Form
- Publish Forms

Creating Web Enabled InfoPath Forms

In this chapter, you will learn how to create web enabled InfoPath forms. InfoPath is an information gathering tool and is used to create XML forms. People have always been trying to come up with a standard for creating electronic forms. Before InfoPath, companies created electronic forms in different formats. InfoPath has helped companies to adhere to a standard. Coding XML manually may not be an interesting job but creating XML forms with InfoPath has really made this job quite easier.

All clients must have InfoPath installed on their machines in order to open the forms. It could be a problem for smaller companies to provide InfoPath client to all the users. What can be the solution then? Solution is to create web enabled forms that can be opened in web browser without having to install InfoPath client on user machines.

Following is a step-by-step guide that will show you how to create and deploy a web enabled form:

1. Open InfoPath (Start > All Programs > Microsoft Office 2013 > InfoPath Designer 2013)

2. Under **Popular Form Templates**, double-click **SharePoint Form Library**.

Click to add title

Click to add heading			
Add label	Add control	Add label	Add control
Add label	Add control	Add label	Add control
Add label	Add control	Add label	Add control

Click to add heading	
CLICK TO ADD SUBHEADING	
Add label	Add control
Add label	Add control
Add label	Add control

Figure 5-1: *New Form*

Figure 5-1 shows how the new form looks.

3. Change the form label. Click on the words **Click to add title** and write **User Registration**.

4. Change section heading. Click on the words **Click to add heading** in the first section and write **User Information**.

5. Highlight second section and delete it.

6. Now it's time to add labels and controls in the form. Add following labels and controls in the form. To add a label, click on **Add label** text and enter the label name. To add a control, click **Add control** words and then click control name to add it there. Controls are located under the **HOME** tab (in **Controls** section). Enter first label called **First Name**. Click **Add label** in the first cell and write **First Name** there. In the second column adjacent to the first cell, click **Add control** and then click **Text Box** in the menu. This will add **Text Box** control in the cell.

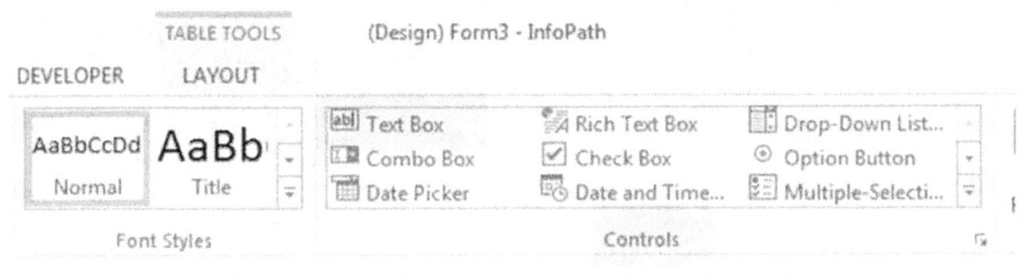

Figure 5-2: *InfoPath controls*

Right-click the text box and select select **Text Box Properties…**. In the **Field Name**, enter **FirstName** and click **OK**. You can also rename control by selecting the respective node on the right side and changing the field name in the properties.

Here is the list of all labels and controls.

Label	Control Name	Control Properties
First Name	FirstName	Under **Data** tab, change the control name to **FirstName**. Under **Display** tab, uncheck **Enable spelling checker** and **Enable AutoComplete**.
Last Name	LastName	Under **Data** tab, change the control name to **LastName**. Under **Display** tab, uncheck **Enable spelling checker** and **Enable AutoComplete**.
Address	Address	Under **Data** tab, change the control name to **Address**. Under **Display** tab, uncheck **Enable spelling checker** and

		Enable AutoComplete. Check **Multi-line** option.
City	City	Under **Data** tab, change the control name to **City**. Under **Display** tab, uncheck **Enable spelling checker** and **Enable AutoComplete**.
State	State	Under **Data** tab, change the control name to **State**. Under **Display** tab, uncheck **Enable spelling checker** and **Enable AutoComplete**.
Zip	Zip	Under **Data** tab, change the control name to **Zip**. Under **Display** tab, uncheck **Enable spelling checker** and **Enable AutoComplete**.
Phone	Phone	Under **Data** tab, change the control name to **Phone**. Under **Display** tab, uncheck **Enable spelling checker** and **Enable AutoComplete**.
Email	Email	Under **Data** tab, change the control name to **Email**.

		Under **Display** tab, uncheck **Enable spelling checker** and **Enable AutoComplete**.

Table 5-1: *Layout Controls*

7. After adding the above controls, there are two more controls to be added yet. Add two buttons at the bottom of the form. We will change button settings later.

User Registration

Click to add heading

First Name		Last Name	
Address		City	
State		Zip	
Phone		Email	

Figure 5-3: *Complete Form*

One of the buttons will submit the form to a SharePoint library and the other button will close the form. To submit form to a library, you need to create a library in SharePoint first.

8. Go to your SharePoint site. You will create a library at the root site. Open SharePoint site. On the home page, click third image button to add a new library.

Figure 5-4: *Add new library*

9. Under **Apps you can add**, click on **Form Library** icon.

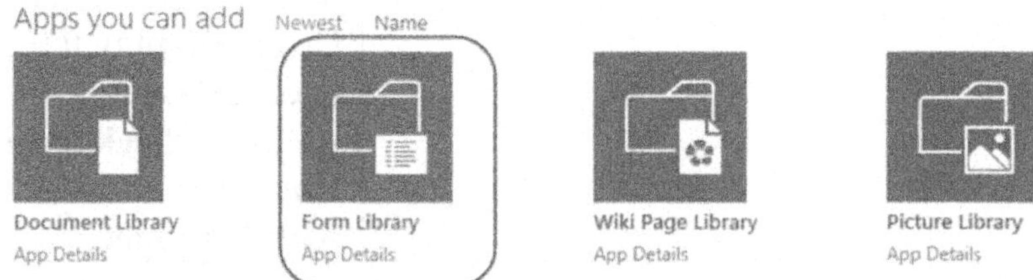

Figure 5-5: *Click Form Library Icon*

10. Enter **User Registration** in the **Name** box and click **Create** button.

11. Now you have created the library, go back to the form. Click on **Manage Data Connections…** link on the right side (at the bottom).

Figure 5-6: *Manage Data Connections*

12. Click **Add…** button.

13. Select **Submit data** option and click **Next**.

14. Select **To a document library on a SharePoint site** and click **Next**.

15. Enter the document library path in the **Document Library** box. For example, "http://yourservername/UserRegistration".

16. Click function (fx) button to create file name.

17. Click on **Insert Function** button. From the **Functions**, select **concat** and hit **OK**.

18. The file name will comprise of user's first name and last name. You will concatenate both fields to create the file name. Double-click first field to insert first name and select **FirstName** from the fields. Click **OK**. Double-click second field and select **LastName**. Click **OK**. Remove third field from the concat function, that is not needed. Click **OK**.

19. Check **Allow overwrite if file exists** and click **Next**.

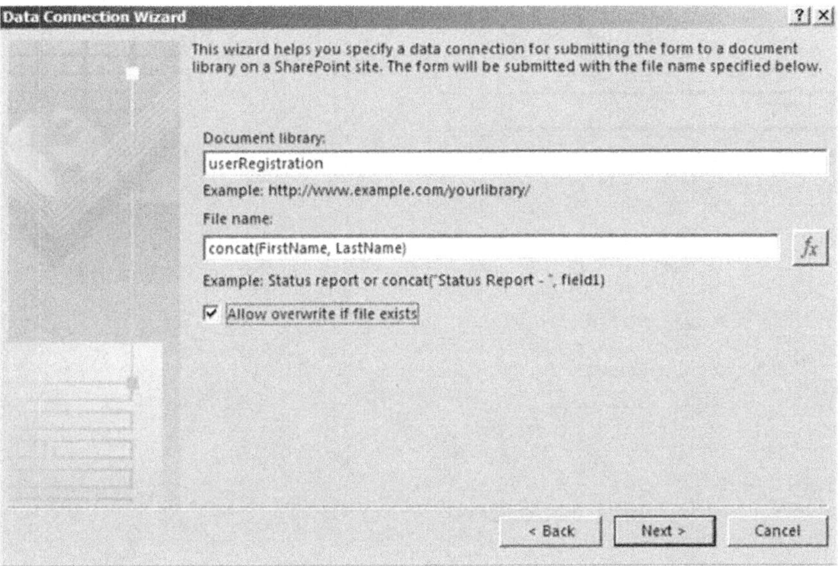

Figure 5-7: *Data Connection Wizard*

20. Click **Finish**. Click **Close**.

21. Right-click first button on the form and select **Button Properties…**. From the **Action** drop down, select **Submit**.

22. Click **Submit Options…**.
23. Check **Allow users to submit this form**.From the first drop down, select **SharePoint document library**.

24. From the second drop down, select the connection that you created for data submission. The default name of the connection was **SharePoint Library Submit**. Select this in the drop down. Click **OK**.

25. Right-click second button and select **Button Properties....** By default, **Action** drop down will have **Rules and Custom Code** selected. Keep that option selected. Change **Label** to **Close**. Change **ID** to **btnClose**. Click **OK**.

26. **Rules** pane will automatically open on the right side. If that does not happen, click **Close** button and then click **Manage Rules** button in the ribbon. That will open rules pane. Click **Add** button in the pane to add new action and select **Close the form**.

Figure 5-8: *Add Rule*

27. **Rule Details** dialog box will open. Make sure **Action** drop down has **Close the form** option selected. Click **OK**.

28. Save the form at this stage. Now the pending work is to change submit options. From the **File** menu, click **Info** and then click **Form Options** button.

29. Select **Web Browser** category. Uncheck **Show InfoPath commands in Ribbon or toolbar**.
30. Select **Property Promotion** category. Click **Add** and select **FirstName**. Similarly, add all other fields one by one. These promoted fields will be available to be added to the library view.

31. Select **Security and Trust** category. Uncheck **Automatically determine security level (recommended).** Select **Full Trust** option.

32. Select **Compatibility** category. Make sure **Form type** has **Web Browser Form**. In **Server validation** section, enter "http://yourservername/_vti_bin/FormsServices.asmx". Replace **yourservername** in the URL with your actual server name.

33. Click **OK**.

34. It's time to publish the form now. From **File** menu, select **Publish** and then click **SharePoint Library**.

35. Enter your server URL and click **Next**.

36. Make sure **Enable this form to be filled out by using a browser** is selected. Click **Next**.

37. Browser to the folder where you want to save the published form. Enter **UserRegistration** in the file name box. Click **Save**.

38. Click **Next.**

39. Click **Next**.

40. Click **Publish**.

41. Click **Close.**

Uploading the Published Form

42. There are two ways to upload the published web form to SharePoint: Using the interface provided by SharePoint and using PowerShell. You will use the first

option in this exercise. Open **SharePoint Central Administration** site by selecting Start > All Programs > Microsoft SharePoint 2013 Products > SharePoint 2013 Central Administration.

43. Select **General Application Settings** and click **Upload form template**.

Figure 5-9: *Upload form template*

44. Click **Browse** button to browse to the folder where form was published and select the published form. Click **Upload** button. You will see a message telling you that form has been uploaded. Click **OK**.

45. Still on the same page (Manage Form Templates), right-click the newly uploaded form and select **Activate to a site collection**. Make sure correct site collection is selected in the **Activation Location** section and click **OK**. In case you get an error, try activating it from the site collection features in the SharePoint site. Here is how you do it. Go to SharePoint site. Click **Site Contents** link on the left side. Click **Settings** link. It will be under the search box. In **Site Collection Administration**

section, click **Site collection features** link. Locate **UserRegistration** text and click **Activate** button in the same row.

Figure 5-10: *Activated Feature*

Preparing Forms Library

46. Go to the **UserRegistration** library. Click **LIBRARY** tab at the top.

47. Click **Library Settings** button in the ribbon.

48. Click **Advanced Settings** link.

49. Set **Allow management of content types?** to **Yes**.

50. Set **Make 'New Folder' command available?** to **No**.

51. Keep all other default values and click **OK**.

52. On the library settings page, in **Content Types** section, click **Add from existing site content types** link.

53. From the list of available site content types, select **UserRegistration** and click **Add** button. This will add the content type to the library. Click **OK**.

54. Click **Change new button order and default content type** link. Uncheck **Form** option and click **OK**.

55. Go back to the main library page and click the new document link. The form will open in browser. Fill it out and click Submit. Also try to the Close button. This is how the final form looks like:

User Registration

User Information			
First Name	John	Last Name	Doe
Address	123 Calvert St	City	Baltimore
State	MD	Zip	21889
Phone	301-333-2344	Email	jdoe@baltimore.com

Submit Close

Figure 5-11: *Completed form ready for submission*

After the form is submitted, open it again by clicking the form name and then close it by hitting the **Close** button.

This was a simple form used to demonstrate the creation and publishing process. You have seen the publishing method. Now you can design simple forms as well as complicated one. The publishing method does not change. Only form layout and design will change. For complicated forms, programming can be done in Visual Studio.

Summary

This chapter explained the basics of creating electronic forms and publishing them in SharePoint. After finishing this chapter, you should have a firm understanding of InfoPath web forms and the publishing procedure. Electronic forms are crucial to your business needs. Paper documents can be converted into electronic forms and distributed through SharePoint sites. It makes managing forms easier and hassle-free.

Chapter 6

Web Part Development

In this chapter:

- What are web parts?
- ASP.NET web parts
- SharePoint web parts
- Developing connectable web parts

Web parts are components or in technical terms server side controls that run inside SharePoint pages. Web parts are building blocks of pages that appear on a SharePoint site. They run inside the context of SharePoint site pages. They are used to customize the user interface and to edit the content of the page. There are several web parts that are available out of the box. There are two types of web parts: ASP.NET web parts and SharePoint web parts. They both serve same purpose but the difference is technical. ASP.NET web parts use the System.Web.UI.WebControls.WebParts.WebPart base class while SharePoint web parts use the Microsoft.SharePoint.WebPartPages.WebPart base class. Both types are supported but ASP.NET web parts are recommended for the new projects.

ASP.NET Web Parts

The ASP.NET web part infrastructure is based on a **WebPartManager** class. WebPartManager class manages the lifetime of web part instance at run time. Each ASP.NET page must contains exactly one **WebPartManager** object and one or more **WebPartZone** objects. **WebPartManager** object tracks which web parts have been added to each particular zone and stores and retrieves data about how each web part has been customized and personalized. **WebPartZone** is the object into which web part is placed. The **SPWebPartManager** and **WebPartZone** controls manage the serialization of data associated with web parts into the appropriate content database. Both these controls are required on the page that contain web parts and for this reason you cannot simply copy ASP.NET page into SharePoint. To move ASP.NET web parts from ASP application to SharePoint application, you have to export them as .webpart files and import them into SharePoint.

SharePoint Web Parts

The web part infrastructure uses many of the controls in the ASP.NET web part control set and introduces several of its own controls that inherit from base classes supplied by ASP.NET web parts control set.

Site pages in SharePoint do not use the standard ASP.NET **WebPartManager** class. Instead, they use the SharePoint specific **SPWebPartManager** class that inherits from the ASP.NET **WebPartManager**.

Web Part Definitions

A web part definition (.webpart) file is a simple XML file that contains property settings for a single web part. To import web part into SharePoint, upload .webpart file to the web parts gallery and add the web part to the page. .webpart file is also used in the web part deployment packages.

SafeControl Entries

To prevent untrusted users from adding server side code in pages, SharePoint provides a SafeControls list. This is a list of approved controls and web parts specific to your SharePoint site that you have designated as safe for invocation on any page. This list is contained in the web.config file of the SharePoint web application.

Full Trust Security

The assemblies in a farm solution in previous versions of SharePoint could be deployed with Custom Access Security (CAS) policies. This method of securing server-side code was quite useful because it stopped malicious code from reaching the server infrastructure. CAS polices were used in SharePoint 2003, 2007 and 2010. Such policies are now ignored in SharePoint 2013, all assemblies in far solutions in SharePoint 2013 run with full trust. If you open web application's web.config, you will notice that <trust> has a new attribute called **LegacyCasModel** and it is set to **true**.

<trust level="Full" originUrl="" legacyCasModel="true" />

This attribute is set to true to enable previously written applications with CAS policies to still function. Assemblies can still be deployed to either GAC or Bin folder but deploying them in Bin does not require to setup custom permissions like before.

Custom Web Parts

Custom web parts provide a method to create user interface elements that support both customization and personalization. Customized changes are seen by all the site users. Customization means that individual users can also further personalize web parts page by adding, removing and configuring web parts. Personalization means that changes are done only for the user who made them. SharePoint web part infrastructure is built on top of the ASP.NET web parts control set, therefore developers can reuse their knowledge of ASP.NET programming to create robust custom web parts. Custom web parts extend out-of-the-box SharePoint functionality. The advantages of using web parts are as following:

You can create custom properties that you can display and modify in the user interface.

- You can improve performance because custom web part runs faster than a script.
- You can implement functionality without disclosing the source code.
- Built-in web parts enable all users who have appropriate permissions to change the settings and content. With custom web part, developer can determine the content and properties to display to users, regardless of their permissions.
- You can make your web part connectable. This allows web parts to provide or access data from other connectable web parts.
- You can interact with object models and create complex features using the object models.
- You can use built-in cache tools. You can control the cache usage.
- You benefit from rich development environment. You have excellent debugging features in Visual Studio.

- You can create a base class that other web parts can use. It makes development easier.
- Web parts are compatible with a broader range of web browsers.

Connectable Web Parts

The web part infrastructure provides a standardized set of interfaces called connection interfaces that allow web parts to exchange information with another at run time. For example, one web part can provide (send) a row of data to any other web part that can consume (receive) that row. This is modular approach of implementing complex functionality. Instead of forcefully pushing all functionality into one web part, you can create two web parts. One implements the interface and provides data. The other consumes the data and focuses on the presentation of data. The web part that provides data is called **Provider** web part. The web part that consumes data is called **Consumer** web part. These web parts can be developed by entirely different developers or companies to communicate with one another. This enables users to build sophisticated combinations of web parts through a simple menu-driven user interface.

Developing Connectable Web Parts

In this section, you will learn two things, developing web parts and developing connectable web parts. Connectable web parts are like ordinary web parts but they have this extra capability of providing data for consumption to other web parts. You will create connectable web parts from scratch using .NET. In the next section, you will learn to use InfoPath web parts. InfoPath web part comes with SharePoint out of the box and is used to host InfoPath forms. If InfoPath form is configured correctly or in other words, if its fields have been exposed as parameters, it can be used in InfoPath web part.

In this section we will create .NET web parts from scratch and then deploy them in SharePoint. We will create two web parts, one provider and one consumer. You can download complete application/code from the book's website.

Please note that web part deployment package (WSP) is in the following folder:

ConnectableWebParts.zip -> ConnectableWebParts\ConnectableWebParts\bin\Debug\ConnectableWebParts.wsp

So, let's begin. Here are the steps:

1. Open Visual Studio. You can develop this project in any version of Visual Studio (2010, 2012, or 2013). The look and feel of your environment may be slightly different that the screenshots. Click **New Project**.

2. Under **Installed Templates**, expand **SharePoint** node and select **2010** (assuming you are using Visual Studio 2010 otherwise select the appropriate folder for your environment). Select **Empty SharePoint Project** and enter **ConnectableWebParts** in the **Name** box and click **OK**. You can change the Location if you want otherwise keep the default location selected.

3. Select a site in **What local site do you want to use for debugging**. Select **Deploy as a farm solution** option and click **Finish**. This will create a new project for you.

4. Right-click project name and select **Add.** Select **New Item**. In **Installed Templates**, Under **Visual C#,** select **Code** and then select **Interface**. Enter **IParams.cs** in the **Name** box and click **Add**.

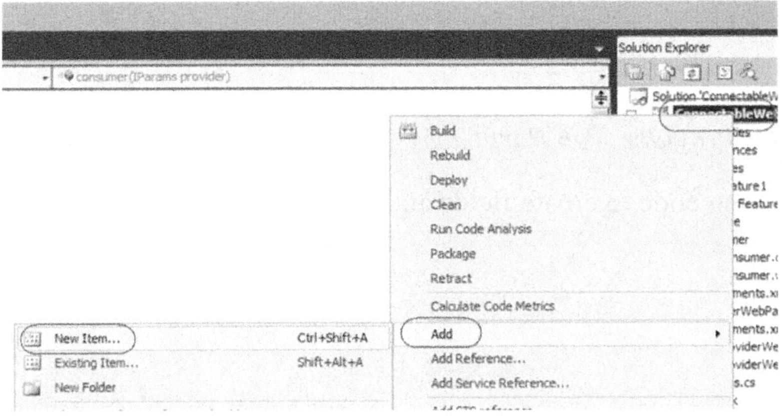

Figure 6-1: *Add new item*

5. In IParams.cs code, add public before interface IParams and then add string parameters. Code will look like the following:

```
namespace ConnectableWebParts
{
    public interface IParams
    {
        string sFirstName { get; set; }
        string sLastName { get; set; }
        string sAddress { get; set; }
        string sCity { get; set; }
        string sState { get; set; }
        string sZip { get; set; }
        string sCountry { get; set; }
    }
}
```

Figure 6-2: *Interface*

6. Now we will create a provider web part. Right-click project name. Select **Add**. Select **New Item**. In **Installed Templates**, select **Visual C#**. Select **SharePoint**. Select **2010 (or 2012 if are using VS 2012)**. Select **Web Part**. Enter **ProviderWebPart** in **Name** box and click **Add**.

7. **ProviderWebPart** class inherits from **WebPart**. Add **IParams** after **WebPart**.

```
namespace ConnectableWebParts.ProviderWebPart
{
    [ToolboxItemAttribute(false)]
    public class ProviderWebPart : WebPart,IParams
    {
        private TextBox firstName;
        private TextBox lastName;
```

Figure 6-3: *Provider Web Part Class*

8. Add following code to create fields and Submit button.

```
[ToolboxItemAttribute(false)]
public class ProviderWebPart : WebPart,IParams
{
    private TextBox firstName;
    private TextBox lastName;
    private TextBox address;
    private TextBox city;
    private TextBox state;
    private TextBox zip;
    private TextBox country;
    private String inputFirstName = string.Empty;
    private String inputLastName = string.Empty;
    private String inputAddress = string.Empty;
    private String inputCity = string.Empty;
    private String inputState = string.Empty;
    private String inputZip = string.Empty;
    private String inputCountry = string.Empty;

    private Button _button;
```

Figure 6-4: *Properties*

9. Add get and set methods.

```
[Personalizable]
public string sFirstName
{
    get
    {
        return inputFirstName;
    }
    set
    {
        inputFirstName = value;
    }
}

[Personalizable]
public string sLastName
{
    get
    {
        return inputLastName;
    }
    set
    {
        inputLastName = value;
```

```
    }
}

[Personalizable]
public string sAddress
{
    get
    {
        return inputAddress;
    }
    set
    {
        inputAddress = value;
    }
}

[Personalizable]
public string sCity
{
    get
    {
        return inputCity;
    }
    set
    {
        inputCity = value;
    }
}

[Personalizable]
public string sState
{
    get
    {
        return inputState;
    }
    set
    {
        inputState = value;
    }
}
```

```
[Personalizable]
public string sZip
{
    get
    {
        return inputZip;
    }
    set
    {
        inputZip = value;
    }
}

[Personalizable]
public string sCountry
{
    get
    {
        return inputCountry;
    }
    set
    {
        inputCountry = value;
    }
}
```

Listing 6-1: *Add Get and Set Methods*

10. Modify **CreateChildControls()** method.

```
protected override void CreateChildControls()
{
    Controls.Clear();
    firstName = new TextBox();
    lastName = new TextBox();
    address = new TextBox();
    city = new TextBox();
    state = new TextBox();
    zip = new TextBox();
    country = new TextBox();

    Controls.Add(firstName);
    Controls.Add(lastName);
    Controls.Add(address);
    Controls.Add(city);
    Controls.Add(state);
    Controls.Add(zip);
    Controls.Add(country);

    _button = new Button();
    _button.Text = "Submit";
    Controls.Add(_button);
    _button.Click +=new EventHandler(_button_Click);

}
```

Figure 6-5: *CreateChildControls()*

11. Handle Click event.

```
void _button_Click(object sender, EventArgs e)
{
    if (firstName.Text != string.Empty)
    {
        sFirstName = firstName.Text;
        firstName.Text = string.Empty;
    }

    if (lastName.Text != string.Empty)
    {
        sLastName = lastName.Text;
        lastName.Text = string.Empty;
    }

    if (address.Text != string.Empty)
    {
        sAddress = address.Text;
        address.Text = string.Empty;
    }
```

```
    if (city.Text != string.Empty)
    {
        sCity = city.Text;
        city.Text = string.Empty;
    }

    if (state.Text != string.Empty)
    {
        sState= state.Text;
        state.Text = string.Empty;
    }

    if (zip.Text != string.Empty)
    {
        sZip = zip.Text;
        zip.Text = string.Empty;
    }

    if (country.Text != string.Empty)
    {
        sCountry = country.Text;
        country.Text = string.Empty;
    }
}
```

Listing 6-2: *Handle click event*

12. Create a method to return the iParams object.

```
[ConnectionProvider("Provider", "Provider")]
public IParams provider()
{
    return this;
}
```

Figure 6-6: *Connection Provider*

13. Now let's create a **Consumer** web part. Just like you added a provider web part above, add a new web part and name it **Consumer**.

14. Open **Consumer.cs** and add code to create objects for the provider.

```
[ToolboxItemAttribute(false)]
public class Consumer : WebPart
{
    private IParams _provider;
    private Label lblFirstName;
    private Label lblLastName;
    private Label lblAddress;
    private Label lblCity;
    private Label lblState;
    private Label lblZip;
    private Label lblCountry;
```

Figure 6-7: *Consumer web part properties*

15. Add a function that will build rows in a table. We will use a table to show the data retrieved from the provider web part.

```
private static TableRow BuildTableRow()
{
    TableRow row = new TableRow();
    row.Cells.Add(new TableCell());
    row.Cells.Add(new TableCell());
    return row;
}
```

Figure 6-8: *BuildTableRow() Function*

16. Add values to the labels.

```
protected override void OnPreRender(EventArgs e)
{
    EnsureChildControls();
    if (_provider != null)
    {
        lblFirstName.Text = _provider.sFirstName;
        lblLastName.Text = _provider.sLastName;
        lblAddress.Text = _provider.sAddress;
        lblCity.Text = _provider.sCity;
        lblState.Text = _provider.sState;
        lblZip.Text = _provider.sZip;
        lblCountry.Text = _provider.sCountry;
    }

}
```

Figure 6-9: *OnPreRender() Function*

17. Add code in **CreateChildControls()** method.

```
protected override void CreateChildControls()
{
    Controls.Clear();
    Table tbl = new Table();
    TableRow trFirstName = BuildTableRow();
    TableRow trLastName = BuildTableRow();
    TableRow trAddress = BuildTableRow();
    TableRow trCity= BuildTableRow();
    TableRow trState = BuildTableRow();
    TableRow trZip = BuildTableRow();
    TableRow trCountry = BuildTableRow();
    lblFirstName = new Label();
    lblFirstName.Text = "First Name";
    Controls.Add(lblFirstName);
    lblLastName = new Label();
    lblLastName.Text = "Last Name";
    Controls.Add(lblLastName);
    lblAddress = new Label();
    lblAddress.Text = "Address";
    Controls.Add(lblAddress);
    lblCity = new Label();
    lblCity.Text = "City";
    Controls.Add(lblCity);
    lblState = new Label();
    lblState.Text = "State";
```

```
Controls.Add(lblState);
lblZip = new Label();
lblZip.Text = "Zip";
Controls.Add(lblZip);
lblCountry = new Label();
lblCountry.Text = "Country";
Controls.Add(lblCountry);

trFirstName.Cells[0].Text = "First Name";
trFirstName.Cells[1].ForeColor = System.Drawing.Color.Blue;
trFirstName.Cells[1].Font.Bold = false;
trFirstName.Cells[1].Font.Size = FontUnit.Small;
trFirstName.Cells[1].Controls.Add(lblFirstName);

trLastName.Cells[0].Text = "Last Name";
trLastName.Cells[1].ForeColor = System.Drawing.Color.Blue;
trLastName.Cells[1].Font.Bold = false;
trLastName.Cells[1].Font.Size = FontUnit.Small;
trLastName.Cells[1].Controls.Add(lblLastName);

trAddress.Cells[0].Text = "Address";
trAddress.Cells[1].ForeColor = System.Drawing.Color.Blue;
trAddress.Cells[1].Font.Bold = false;
trAddress.Cells[1].Font.Size = FontUnit.Small;
trAddress.Cells[1].Controls.Add(lblAddress);

trCity.Cells[0].Text = "City";
trCity.Cells[1].ForeColor = System.Drawing.Color.Blue;
trCity.Cells[1].Font.Bold = false;
trCity.Cells[1].Font.Size = FontUnit.Small;
trCity.Cells[1].Controls.Add(lblCity);

trState.Cells[0].Text = "State";
trState.Cells[1].ForeColor = System.Drawing.Color.Blue;
trState.Cells[1].Font.Bold = false;
trState.Cells[1].Font.Size = FontUnit.Small;
trState.Cells[1].Controls.Add(lblState);

trZip.Cells[0].Text = "Zip";
trZip.Cells[1].ForeColor = System.Drawing.Color.Blue;
trZip.Cells[1].Font.Bold = false;
trZip.Cells[1].Font.Size = FontUnit.Small;
```

```
        trZip.Cells[1].Controls.Add(lblZip);

        trCountry.Cells[0].Text = "Country";
        trCountry.Cells[1].ForeColor = System.Drawing.Color.Blue;
        trCountry.Cells[1].Font.Bold = false;
        trCountry.Cells[1].Font.Size = FontUnit.Small;
        trCountry.Cells[1].Controls.Add(lblCountry);

        tbl.Rows.AddRange(new TableRow[]{
            trFirstName,
            trLastName,
            trAddress,
            trCity,
            trState,
            trZip,
            trCountry
        });
        this.Controls.Add(tbl);
}
```

Listing 6-2: *CreateChildControls() method*

18. Create a method to assign the IParams object from the provider web part to the private IParams object in the consumer web part.

```
[ConnectionConsumer("Consumer", "Consumer")]
public void consumer(IParams provider)
{
    _provider = provider;
}
```

Figure 6-10: *Connection Consumer*

19. You have completed the code. Now it's time to test the web parts. One way is to use F5 in Visual Studio if you want to debug and test the functionality. Second is to deploy the web part in SharePoint and then test it. I would recommend that you use F5 to deploy the web parts in SharePoint. This will automatically create the WSP for the web parts. Once the WSP has been created, you can deploy it to SharePoint using PowerShell. After the WSP has been created, open PowerShell. Go to **Start > All Programs >**

Microsoft SharePoint 2013 Products and select **SharePoint 2013 Management Shell**.

20. For convenience copy the WSP from the debug folder to a folder that is easily accessible. Go to project folder and then access **bin\debug**. Copy **ConnectableWebParts.wsp** to **C:\Temp** folder (or any other conveniently located folder).

21. From Management Shell command prompt, run following command:

Add-spsolution –literalpath "C:\Temp\ConnectableWebParts.wsp"

```
Administrator: SharePoint 2010 Management Shell
PS C:\Users\Administrator> add-spsolution -literalpath "c:\temp\ConnectableWebPa
rts.wsp"

Name                          SolutionId                             Deployed
----                          ----------                             --------
connectablewebparts.wsp       551c2645-5bce-4486-8602-eb816400dfc1   False
```

Figure 6-11: *Deploy web part*

You will immediately see the result of the deployed. **Deployed** column will show **False**.

22. Now run following command from the Management Shell command prompt:

Install-spsolution –identity connectablewebparts.wsp –webapplication http://yourserver –gacdeployment

Replace http://yourserver with your SharePoint Server (site) URL.

23. Open SharePoint site. Select **Site Actions** and then **Site Content**.

24. Go to **Pages** library.

25. Go to **FILES** tab and click **New Document** button.

26. Enter **Connectable Web Parts** in Title field. Enter **webparts** in **URL Name**. Select **Blank Web Part Page** from **Page Layout**. Click **Create**.

27. Open the newly created page. Edit it. In the **Header** section, click **Add a Web Part** link. In **Categories** select **Custom** and in **Parts** select

ProviderWebPart. In **Center** section, click **Add a Web part** link. In **Categories** select **Custom** and in **Parts** select **Consumer.**

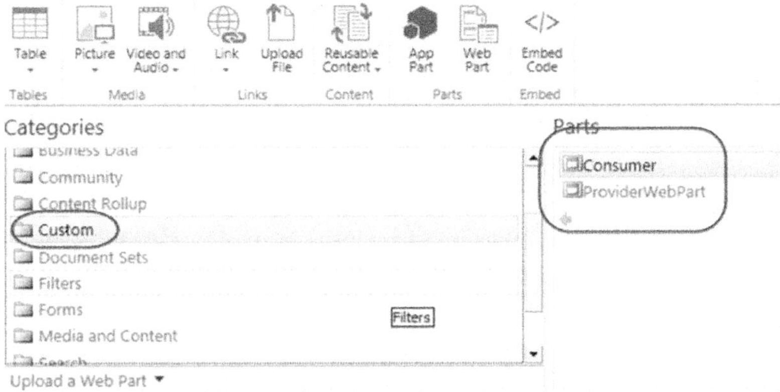

Figure 6-12: *Add web parts*

29. Hover mouse over **Consumer** web part heading. A small arrow will appear at the end of the row that contains the heading. Click the arrow, a menu will appear. Select **Connections > Get Consumer From > ProviderWebPart**.

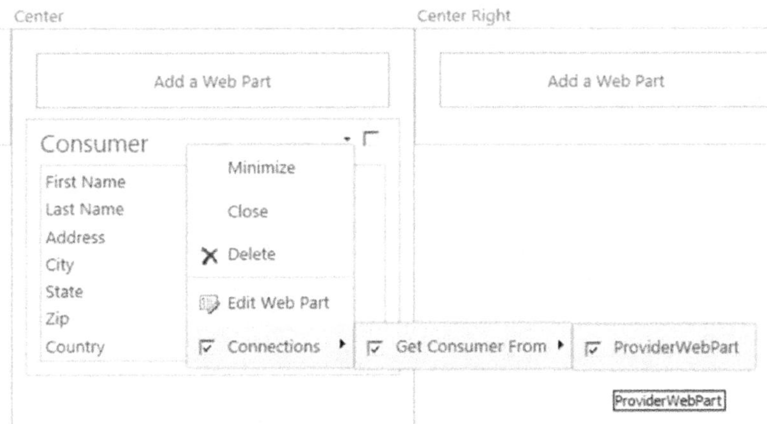

Figure 6-13: *Get Consumer From*

30. Click **Save** button in the ribbon.

31. Add data in the provider web part and hit **Submit**. Before submitting data, this is how the web parts will look:

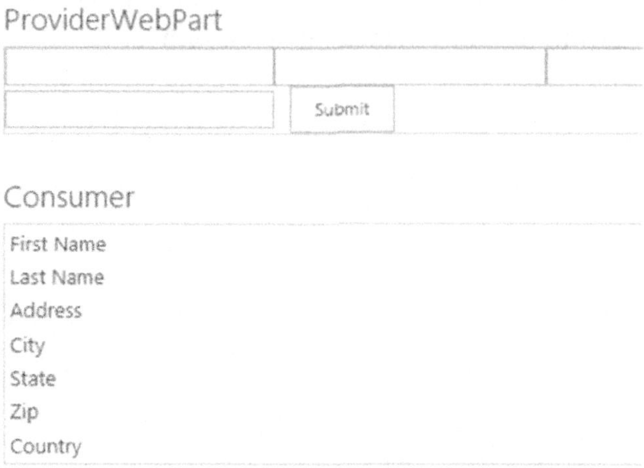

ProviderWebPart

Submit

Consumer

First Name
Last Name
Address
City
State
Zip
Country

Figure 6-14: *Provider and Consumer Web Parts ready for use*

After you submit data, Consumer web part will show values from the provider web part. This is how it will look:

ProviderWebPart

Submit

Consumer

First Name	Saifullah
Last Name	Shafiq
Address	Rooks Ct
City	Frederick
State	MD
Zip	21703
Country	USA

Figure 6-15: *Provider and Consumer Web Parts after Submission*

Next you will learn to use InfoPath web part.

InfoPath Form Web Part

I assume you have read chapter on InfoPath forms. In that chapter, you created and published a web-enabled InfoPath form. Now, you will learn to create an InfoPath form that can be used in **InfoPath Form Web Part**. As explained in the previous section, web parts offer an easy way of customizing user interfaces. Instead of sending the user to a separate page for filling out data, you can display the form on the main page using a web part. Developing forms in InfoPath are easier than developing them in Visual Studio. Creating cool form interface is easier in InfoPath 3and you have the power of Visual Studio to add power functionality in the form. Once you have form, you can use it as stand-alone or embed on any SharePoint page using the web part.

This is a very exciting topic. Have you ever created or attempted to create connected web parts in older versions of SharePoint, for example, version 2003 or 2007? You needed to have very good .NET programming skills in order to do that and also required good understanding of SharePoint object model. In SharePoint 2013 (and in SharePoint 2010 before this version), creating connected web parts is considerably easy. You still need to do .NET programming if you want to create connectable web parts in .NET but SharePoint offers another new interesting way of creating these web parts and that is by using InfoPath. SharePoint 2010 had introduced a new kind of web part called InfoPath Form Web Part. This web part is still available in SharePoint 2013. This web part allows you to load InfoPath form. Therefore, you don't have to do any programming as far as web part is concerned. You just load the InfoPath form and form functionality can then be accessed through the web part. InfoPath form allows you to promote fields to be used in connectable web parts. Just like you promote fields to be viewable in the SharePoint library view, you can now promote fields to be used in web parts. Below you will create a browser-enabled InfoPath form and promote its fields to be used in connectable web parts. Later you will see how this form can be used in **InfoPath Form Web Part** and how you can connect one web part to the other.

So the first goal is to create an InfoPath form that will be used as a connectable web part later. Please note that if you have already created InfoPath form and just want to see how it works as a connectable web part, please skip the sections below that show how to setup the environment and create InfoPath form. You can directly jump to the section that shows how to use **InfoPath Form Web Part** to show the InfoPath form. Important step that you should not miss is InfoPath publishing because this is where you add fields you want to be exposed in the connectable web part. I will skip some screenshots below. If you get stuck, please go back to the chapter on InfoPath forms for reference and details.

Create Form Library

1. First you will setup a form library in SharePoint. You will publish form to this library. Open SharePoint site. You will create a library at the root site. Open SharePoint site. On the home page, click third image button to add a new library.

Figure 6-16: *Create form library*

2. Under **Apps you can add**, click on **Form Library** icon.

3. Enter **UserRegistrationLib** in the Name box and click **Create**.

4. The library has been created. Now we will create the form. We will return to this library for some more settings after the form has been created and deployed.

Create Form

5. To create the form, open InfoPath designer (**All Programs > Microsoft Office 2013 > InfoPath Designer 2013**).

6. Click **File** tab and this time select (double-click) **SharePoint Form Library.** InfoPath designer will show a new form.

7. To edit this form, click **Click to add title** and enter **User Registration Form**.

8. Click first heading **Click to add heading** and enter **Personal Information**.

9. To change field name, click it and enter a new name. Enter following field names/labels in the first section:
    a. First Name
    b. Last Name
    c. Address
    d. City
    e. State
    f. Zip
    g. Country
    h. Phone
    i. Email
    j. Web

10. Click second heading **Click to add heading** and enter **Professional Skills**.

11. Add following field names/labels in this second section:
    a. Description
    b. Skills
    c. Experience (Number of years)

User Registration Form

Personal Information			
First Name		Last Name	
Address		City	
State		Zip	
Country		Phone	
Email		Web	

Professional Skills	
Description	
Skills	
Experience (Number of years)	

Submit

Figure 6-17: *Form layout*

12. Now, it's time to add controls.

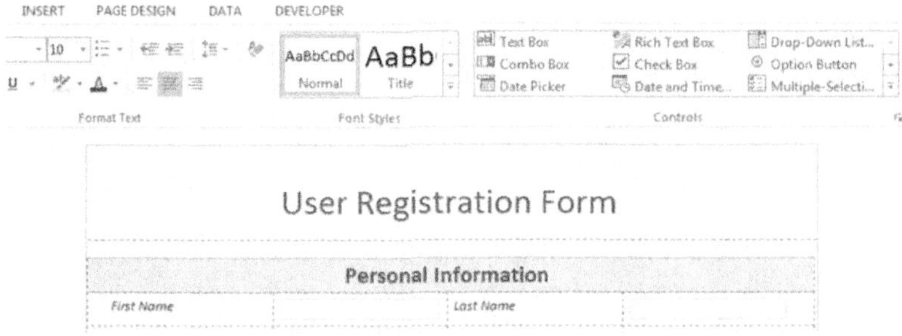

Figure 6-18: *InfoPath controls*

Add following controls from **Controls** section of the ribbon.

Field Name	**Control**	**Properties**
First Name	Text Box	**Data Tab** Field name: FirstName **Display Tab** Uncheck **Enable spelling checker** Uncheck **Enable AutoComplete**
Last Name	Text Box	**Data Tab** Field name: LastName **Display Tab** Uncheck **Enable spelling checker** Uncheck **Enable AutoComplete**
Address	Text Box	**Data Tab** Field name: Address **Display Tab** Uncheck **Enable spelling checker** Uncheck **Enable AutoComplete** Check **Multi-line** **Size** Height: 74 px
City	Text Box	**Data Tab** Field name: City **Display Tab** Uncheck **Enable spelling checker** Uncheck **Enable AutoComplete**
State	Text Box	**Data Tab** Field name: State **Display Tab** Uncheck **Enable spelling checker** Uncheck **Enable AutoComplete**
Zip	Text Box	**Data Tab** Field name: Zip **Display Tab**

		Uncheck **Enable spelling checker** Uncheck **Enable AutoComplete**
Country	Text Box	**Data Tab** Field name: Country **Display Tab** Uncheck **Enable spelling checker** Uncheck **Enable AutoComplete**
Phone	Text Box	**Data Tab** Field name: Phone **Display Tab** Uncheck **Enable spelling checker** Uncheck **Enable AutoComplete**
Email	Text Box	**Data Tab** Field name: Email **Display Tab** Uncheck **Enable spelling checker** Uncheck **Enable AutoComplete**
Web	Text Box	**Data Tab** Field name: Web **Display Tab** Uncheck **Enable spelling checker** Uncheck **Enable AutoComplete**
Description	Text Box	**Data Tab** Field name: Description **Display Tab** Uncheck **Enable spelling checker** Uncheck **Enable AutoComplete** Check **Multi-line** **Size** Height: 90 px

Skills	Text Box	**Data Tab** Field name: Skills **Display Tab** Uncheck **Enable spelling checker** Uncheck **Enable AutoComplete** Check **Multi-line** **Size** Height: 90 px
Experience (Number of years)	Text Box	**Data Tab** Field name: Experience **Display Tab** Uncheck **Enable spelling checker** Uncheck **Enable AutoComplete** Check **Multi-line** **Size** Height: 90 px

Table 6-1: *Form Controls*

13. Add a button control at the bottom. Open button properties. In **Action**, select **Submit**. In **Label**, enter **Submit** and then click **Submit Options**. On **Submit Options** dialog box, check **Allow users to submit this form.** Select **Send form data to a single destination** and select **SharePoint document library** from the drop down. Click **Add** button.

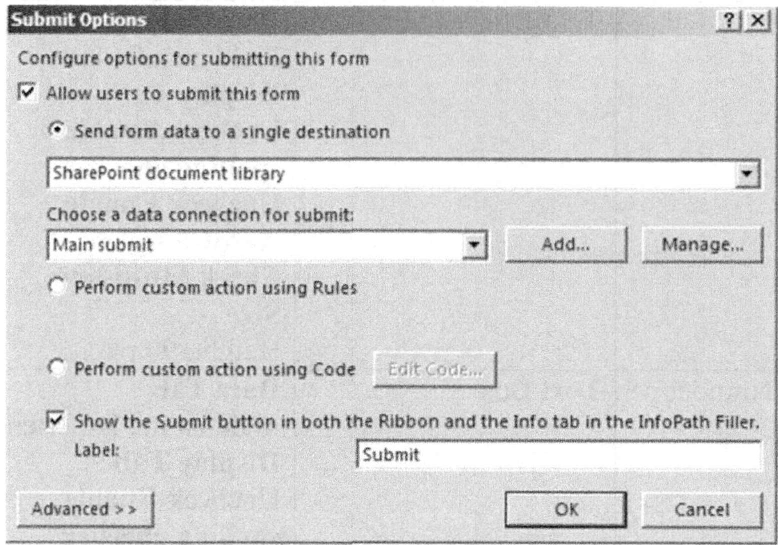

Figure 6-19: *Submit Options*

14. Enter Form library path name in **Document Library**. We created form library in the steps above and named it **UserRegistrationLib**. The path will be something like this http://servername/libname.

15. Click File name button.

Figure 6-20: *File Name*

16. Click **Insert Function** button. Select **concat** from the functions and click **Ok**. Concat function will have three parameters. Delete first parameter and add **"Form-"** in its place. Click on 2nd parameter and then select **Insert Function** button. Select **now** from the functions. Click **OK**. Click **Verify Formula** button to verify it. Click **OK**. This will give a unique name to the form every time you submit the form.

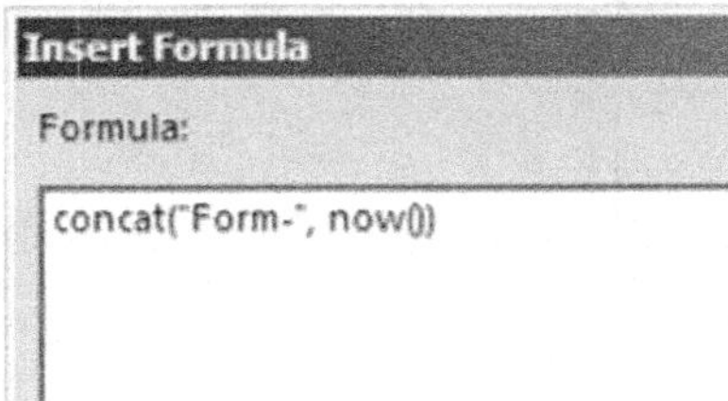

Figure 6-21: *Insert Formula*

17. Click **Next** and click **Finish**.

18. Click **OK** twice to close the dialog boxes.

Publish the Form

19. From the **File** menu, click **Info** and then click **Form Options** button.

20. Select **Web Browser** category. Uncheck **Show InfoPath commands in Ribbon or toolbar**.

21. Select **Property Promotion** category. Click **Add** and select **FirstName**.

Select **Input and output** as parameter type. There are three types of parameters that you can select: **Input, Output and Input and output**. We will use the third one (Input and output) because it will demonstrate the use of both input and output parameters. By selecting **Input and output** type for **FirstName** parameter, we will make this field available as both input and output parameter.

Similarly, add all other fields one by one and for each field the parameter type should be **Input and Output**. These promoted fields will be available to be added to the library view.

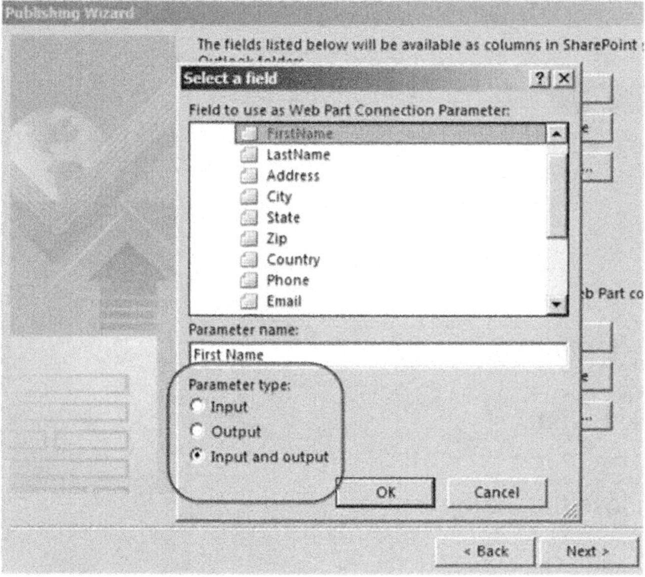

Figure 6-22: *Parameter types*

22. Select **Security and Trust** category. Uncheck **Automatically determine security level (recommended).** Select **Full Trust** option.

23. Select **Compatibility** category. Make sure **Form type** has **Web Browser Form**. In **Server validation** section, enter "http://yourservername/_vti_bin/FormsServices.asmx". Replace **yourservername** in the URL with your actual server name.

24. Click **OK**.

25. It's time to publish the form. From **File** menu, select **Publish** and then click **SharePoint Library**.

26. Enter your server URL and click **Next**.

27. Make sure **Enable this form to be filled out by using a browser** is selected. Click **Next**.

28. Browser to the folder where you want to save the published form. Enter **UserRegistration** in the file name box. Click **Save**.

29. Click **Next.**

30. Click **Next**.

31. Click **Publish**.

32. Click **Close.**

Upload Approved Form to Server

33. Open central administration site. Click **General Application Settings**. Click **Manage form templates** under **InfoPath Forms Services**.

34. Click **Upload form template**.

35. Click **Browse** and select the form that you published (saved on your disk).

36. Click **Upload**.

37. Click **OK**.

38. Right-click published form and select **Activate to a Site Collection**.

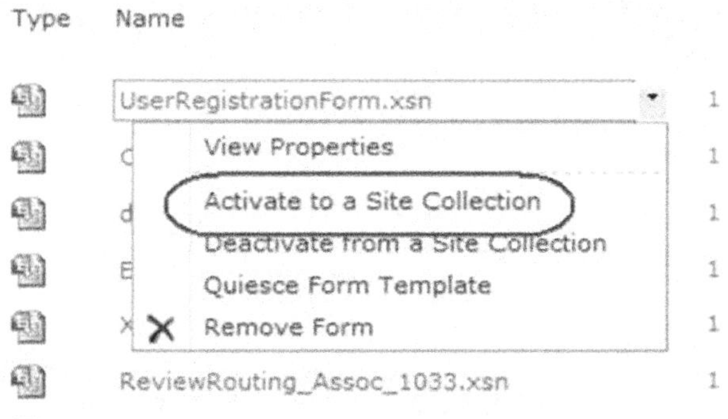

Figure 6-23: *Activate form to a site collection*

39. Make sure correct site collection is selected . You will have to worry about it if you have multiple site collections otherwise the default site collection will be selected automatically. Click **OK**.

Form Library Settings

40. Now we will return to the form library and make some more changes in the settings. Go to the **UserRegistrationLib** library. Click **LIBRARY** tab at the top.

41. Click **Library Settings** button in the ribbon.

42. Click **Advanced Settings** link.

43. Set **Allow management of content types?** to **Yes**.

44. Set **Make 'New Folder' command available?** to **No**.

45. Keep all other default values and click **OK**.

46. On the library settings page, in **Content Types** section, click **Add from existing site content types** link.

47. From the list of available site content types, select **UserRegistrationLib** and click **Add** button. This will add the content type to the library. Click **OK**.

48. Click **Change new button order and default content type** link. Uncheck **Form** option and click **OK**.

Testing the Form

49. Open library. To open click library name in **Libraries**. Click **Add document.**

50. The form will open in browser. Enter some data and click **Submit** button in the form.

User Registration Form

Personal Information

First Name	Billy	Last Name	Bowden
Address	333 James St	City	Frederick
State	MD	Zip	21703
Country	USA	Phone	234 233 3444
Email	billy@bowden.com	Web	billy.com

Professional Skills

Description	Billy is a SharePoint and InfoPath consultant. He has been creating SharePoint portals and InfoPath custom applications since 2004.
Skills	SharePoint, InfoPath, .NET, SQL
Experience (Number of years)	14

Submit

Figure 6-24: *Form in action!*

51. Now next step is to try out this form in the web part. Open SharePoint site. Click **Site Actions** and select **Site Content**.

52. Click **Pages** library. Select **FILES** tab and click **New Document** button. Enter **InfoPathWebParts** in **Title** box and enter **InfoPathWebParts** in **URL Name**. In **Page Layout** select **Blank Web Part page**. Click **Create**.

53. Go to the newly created page. Edit it.

54. Click **Add a Web Part** in the **Header** zone.

55. Select **Forms** in **Categories**. Select **InfoPath Form Web Part** in **Web Parts**. Click **Add**.

56. Click **Click here to open the tool pane.**

57. Select **UserRegistrationLib** or any other library that you want to use and has InfoPath form deployed to. If that library has an InfoPath content type, that will automatically show up in the **Content Type** drop down. Uncheck **Show InfoPath Ribbon or toolbar**. Keep **Send data to connected Web Parts when page** loads checked. Click **Ok**.

58. You will use the same web part twice, one as a provider and the other as a consumer. Remember you can get data from any other provider web part. Similarly we can send data to any other consumer web part. Click **Add a Web Part** in **Footer** zone.

59. Select **Forms** in **Categories**. Select **InfoPath Form Web Part** in **Web Parts**. Click **Add**.

60. Click **Click here to open the tool pane.**

61. Select **UserRegistrationLib** or any other library that you want to use and has InfoPath form deployed to. If that library has an InfoPath content type, that will automatically show up in the **Content Type** drop down. Uncheck **Show InfoPath Ribbon or toolbar**. Keep **Send data to connected Web Parts when page** loads checked. Click **Ok**.

62. Now you have two web parts on this page. To test it, enter a value in **First Name** (in the top web part, the provider).

63. Scroll down. There is a small arrow at the right corner in the row that has the web part name. Click this small arrow. It will open a menu. Select **Connections**. Select **Get Data From**. Select the provider web part.

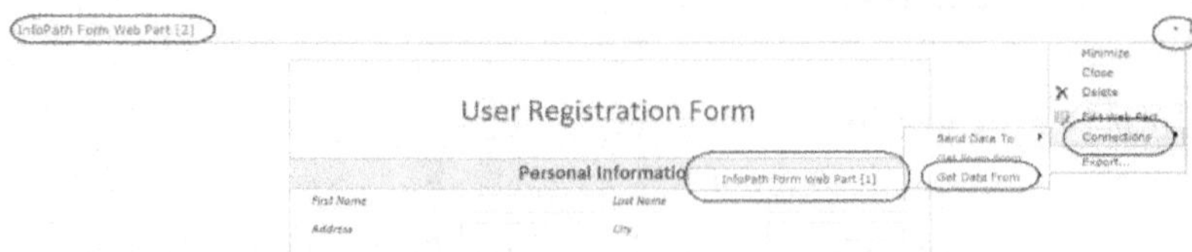

Figure 6-25: *Configure Web Part*

64. **Configure Connection** dialog box will appear. Temporarily allow pop-ups if they are blocked otherwise dialog box will not open. Dialog box will allow you to select provider field name and consumer field name.

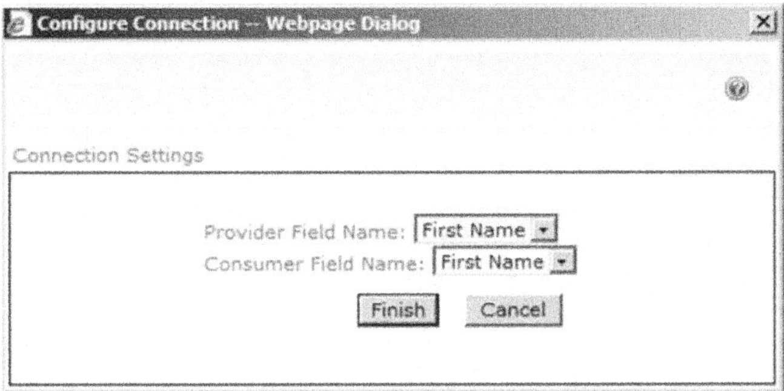

Figure 26: *Provider and Consumer Fields*

65. **First Name** field in the consumer web part will get its value from the **First Name** field of the provider web part. The other way is to open same menu in the provider web part and select **Connections > Send Data To > Consumer Web Part**. This will send the data from **First Name** to the **First Name** field in the consumer web part.

This brings us to the end of this chapter. You have learned to create connectable web parts in .NET and InfoPath. I hope you will be able to take advantage of this knowledge and will apply it in real world scenarios wherever you work.

Summary

Web parts are components or in technical terms server side controls that run inside SharePoint pages. Web parts are building blocks of pages that appear on a SharePoint site. They run inside the context of SharePoint site pages.

The ASP.NET web part infrastructure is based on a WebPartManager class.

The SharePoint web part infrastructure uses many of the controls in the ASP.NET web part control set and introduces several of its own controls that inherit from base classes supplied by ASP.NET web parts control set.

The web part infrastructure provides a standardized set of interfaces called connection interfaces that allow web parts to exchange information with another at run time.

Chapter 7

Setup Alternate URLs

In this chapter:

- Alternate Access Mapping
- Multiple URLs for SharePoint sites
- Creating Site Collection

Alternate Access Mapping

Alternate Access Mapping (AAM) is a feature that allows a website to use different URLs. This is usually needed when you have different zones, for example, one for the Intranet and another for an Extranet. Each web application can be associated with a collection of mappings between internal and public URLs. A public URL is what users type to access the SharePoint site. Internal URLs are in the URL requests that are sent to the SharePoint site. A single public URL can be associated with many internal URLs in multi-server farms.

Configuring AAM is easy. Your default site name is same as your machine name. If your machine name is SP2013, your default site URL will be http://sp2013. If you have a FQDN, your site URL may look like http://sp2013.server.com. In this URL, sp2013 is your SharePoint instance and server.com is your FQDN.

1. Open **Central Admin Site** and click **Application Management**.

2. Under **Web Applications**, click **Configure alternate access mappings**.

3. You will notice you already have http://sp2013 listed there in the **Default** zone. To add new URL in the **Intranet** zone, click **Add Internet URLs**.

4. From the **Alternate Access Mapping Collection** drop down, select correct application that you want to use for the AAM setting and then add URL in the text box labeled **URL protocol, host and port**, for example, http://sp2013.server.com. From the **Zone** dropdown, select **Intranet**.

5. Click **Save**.

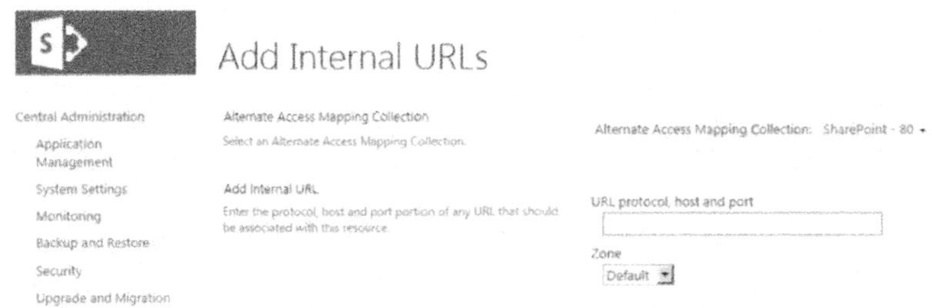

Figure 7-1: *Add Internal URLs*

Creating Alternate URLs for SharePoint Sites

There are different methods in which you can create URLs for SharePoint sites. These methods and some useful tips are briefly discussed below:

1. Add a **DNS** entry. Create a new web application with unique host headers. Create a site collection for this web application. That's it. Just beware that this will not work until you add an entry in the **DNS (A Record)** first. If you add **DNS** entry first and then create web application and site collection then the site works fine but if you create web application and site collection first and then add **DNS** entry, it does not work.

2. Second way is to use alternate access mapping. Choose an existing site and add a new URL. This will not work until you add an **A Record** in the **DNS**. People assume that you don't have to create new web application or extend an existing one. This is wrong assumption. Site does not work until you create new web application using new host header.

3. In **IIS**, you notice that some sites are stopped and when you try to start them you get an error message that tells you that site cannot be started because another application is using same port. One of the scenarios when this happens is when you setup your web application with new host headers and add an entry in the **DNS**. In this case, in IIS, you see that the web site is stopped but when you access it in the browser using the host headers that you had set up, site loads and works fine.

4. You cannot extend a web application using port that already has a site on it. To use same port again, you have to create a new web application and use different host headers. For example, port 80 which is the default port, assume it already has a site associated with it. If you want to create a new web application using port 80 then you have to create a new web application. Note, application at an existing port can be extended but it doesn't work when browsed. It also appears in IIS but when you browse it, you get message that "Internet Explorer cannot display the webpage.". So, this assumption is also wrong that site at an existing port can be extended and then configured using **AAM**. **AAM** cannot be setup for an extended web application, therefore, a different URL cannot be setup for an extended application. New URLs can only be associated with non-extended sites. Extended sites can have only one URL that is defined in the web application's host headers.

Now, you will learn how to create alternate URL for your SharePoint site. By default, you can access SharePoint site by using machine name, for example, if your machine name is MYCOMPANY, you can access SharePoint site using URL http://mycompany (provided you have a site collection set up at that URL). Now, suppose you want to have a meaningful URL for your site, for example, www.mycompany.com OR if you have different departments in your company and you want each department to access their site using their own URL, for example, http://hr.mycompany.com, http://finance.mycompany.com, etc., this can be achieved by configuring DNS. My assumption is that you have admin rights on the machine and you can add DNS role and configure it. If you want to learn how to do that, read the initial sections of this book. Please note that steps below are just a repeat of what has been explained before in the book. For your convenience, the steps are listed again below to not spoil the continuity. In the steps below, you will add **Active Directory Domain Services** and set up a **FQDN**.

1. Go to **All Programs** > **Administrative Tools** and click **Server Manager**.

2. Click **Roles** and then click **Add Roles** link.

3. Click **Next** and then select the second check box **Active Directory Domain Services** and click **Next** >. Click **Next** again and then click **Install**. Installation will start and will take approx. 10 minutes to complete (depending on your computer's speed). Click **Close** once the installation has

completed.

4. Click **Active Directory Domain Services** link. You will see following warning message:

This server is not yet running as a domain controller. Run the Active Directory Domain Services Installation Wizard (dcpromo.exe).

Click the link in this message to install Active Directory Domain Services.

5. Click **Next**. Click **Next** again. Select the second option **Create a new domain in a new forest** and click **Next**.

6. Provide a FQDN, for example, **sp2013.com** and click **Next**. I will use **walisystems.com** as that is my main domain. You will see this domain in the screenshots in this section.

7. Select **Windows Server 2008** from the drop down and click **Next**.

8. Keep **DNS server** checked and click **Next**.

9. You will see a warning message. Select first option **Yes, the computer will use dynamically assigned IP address (not recommended).** If you don't want to assign a static IP to your computer and click **Next**. If you want to make the change, assign static IP and run the wizard again.

10. Next you may see this error:

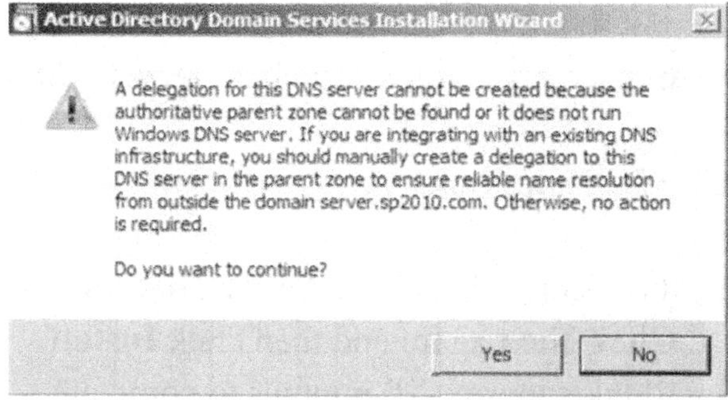

Figure 7-2: *Active Directory Domain Service Installation Wizard*

Click **Yes**.

11. You will be presented with another screen. Keep default values and click **Next**.

12. Enter password for Directory Services Restore Mode Administrator Mode and click **Next**.

13. Review your selections and click **Next**. DNS Installation starts.

14. Click **Finish** after the installation has completed.

15. Click **Restart Now** button to restart your machine.

16. Open **Server Manager** again. Expand **Roles.** Expand **Active Directory Domain Services.** Expand **Active Directory Users and Computers.** You will see your domain listed there.

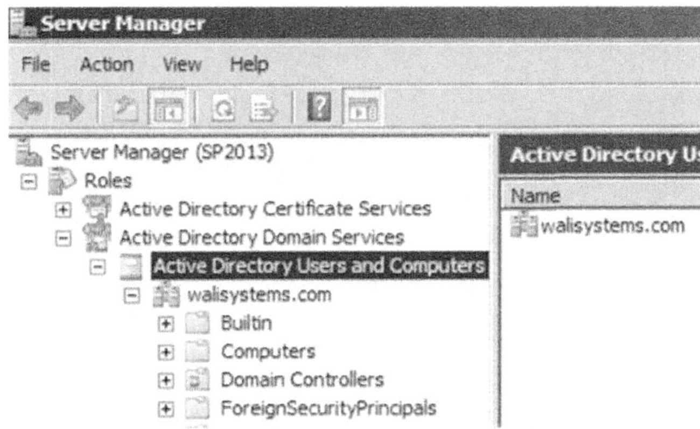

Figure 7-3: *Server Manager*

17. Now you have setup the domain services, you can use multiple URLs to access SharePoint sites. As mentioned earlier, one way to access your SharePoint site is to use the machine name in the URL, for example, http://sp2013 (sp2013 is the machine name). With domain setup on your server, you can now access same site using URL like http://sp2013.walisystems.com. Please note that **walisystems.com** is the **FQDN** that we used while setting up DNS.

Add URLs for Different Sites

In the previous section, you learned how to setup DNS and use FQDN to access sites in SharePoint 2013. In this section, you will learn how to add new URLs for different sites (departments). For example, you can setup a URL for your finance department, HR department, etc.

1. Now you have setup the domain services, you can use multiple URLs to access SharePoint sites. As mentioned in previous section, one way to access your SharePoint site is to use the machine name in the URL, for example, http://sp2013 (sp2013 is the machine name). With domain setup on your server, you can now access same site using URL like http://sp2013.walisystems.com. Please note that **walisystems.com** is the **FQDN**. Now, the fun part starts. We will setup a new URL www.walisystems.com to access your root site collection.

2. Open **Server Manager**. Expand **Roles**. Expand **DNS.** Expand **MACHINENAME**. Expand **Forward Lookup Zones.** Right-click **FQDN** and select **New Host (A or AAAA)….**

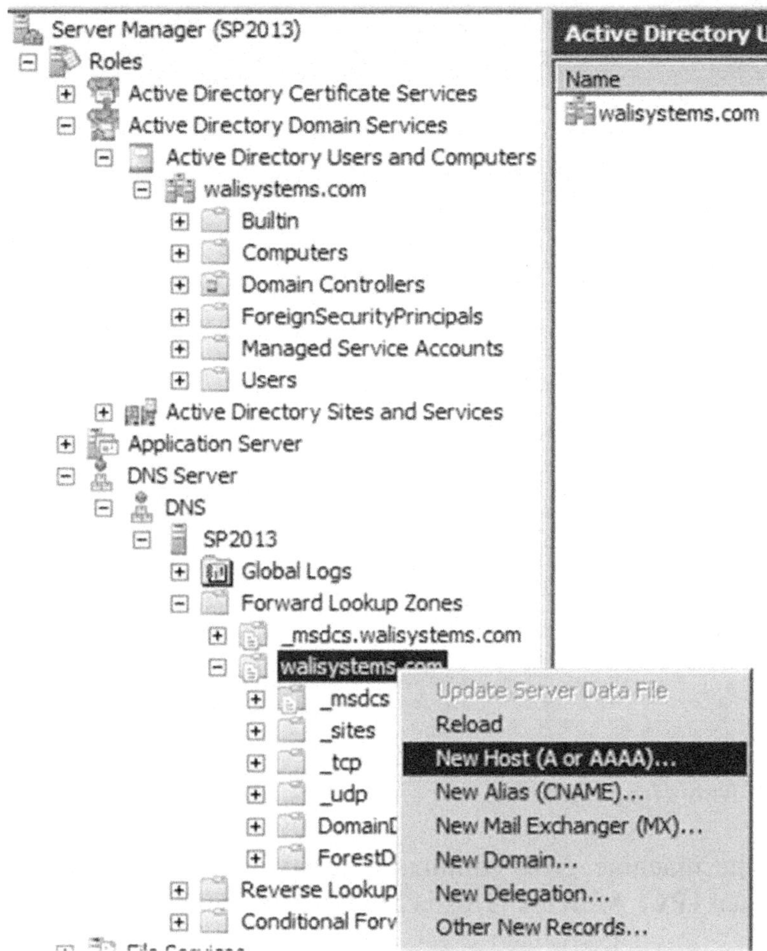

Figure 7-4: *Add a New Host (A Record)*

3. In the **Name** box, enter **www**. In the **IP address**, enter **IP**.

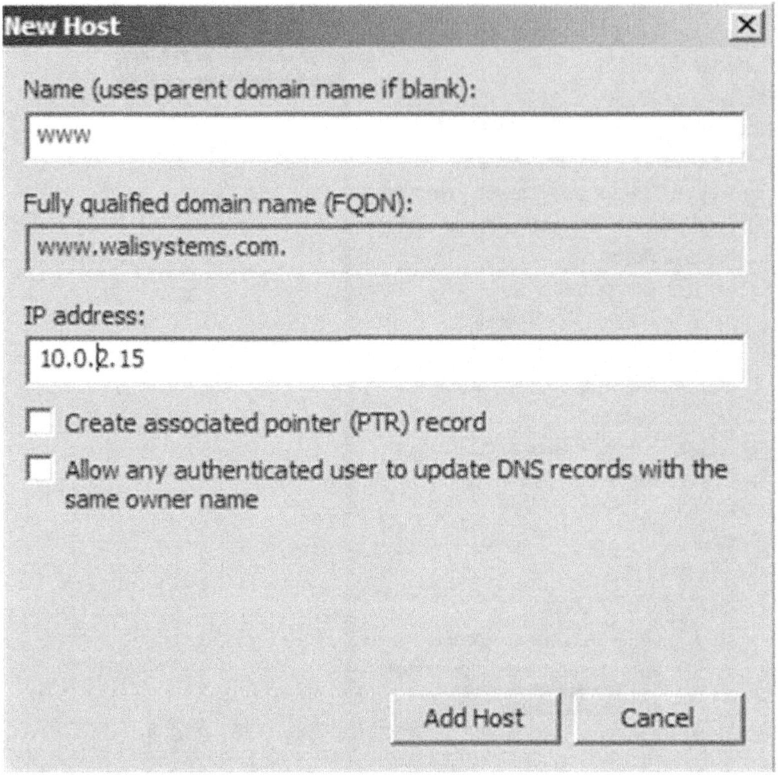

Figure 7-5: *New Host Entry*

To find the **IP** of your machine, go to command prompt and run **IPCONFIG** command. Note down **IPv4 Address**. This is the **IP** of your machine.

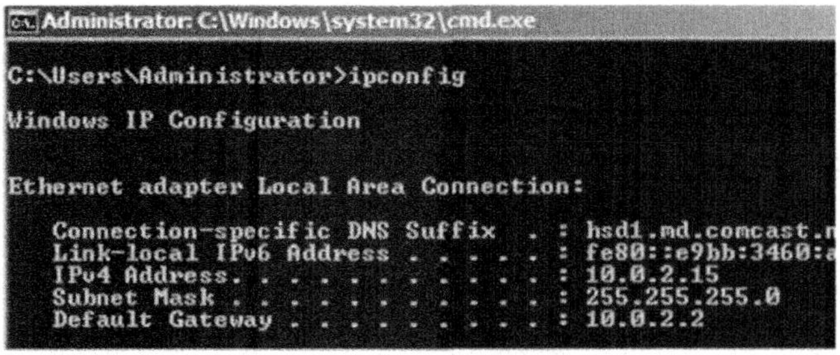

Figure 7-6: *Find IP of your machine*

After entering the **IP**, click **Add Host** button. This will add a new A record.

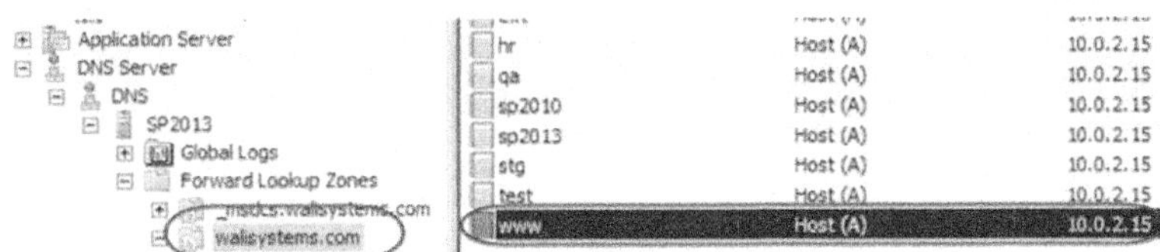

Figure 7-7: *New Host Entry Added*

4. Now, we will create a new web application and site in SharePoint that will use this new URL. Open **Central Administration** site. Click **Application Management.** Click **Manage web applications**.

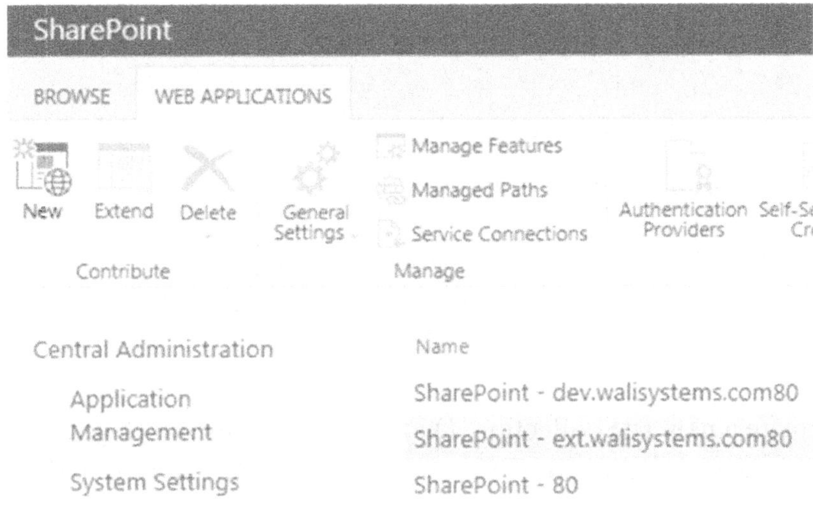

Figure 7-8: *Web Applications*

You will see there are two web applications (if it's a new setup). One is your main site and the other central administration site.

5. Click **New** button in the ribbon to create a new web application. A form will open.

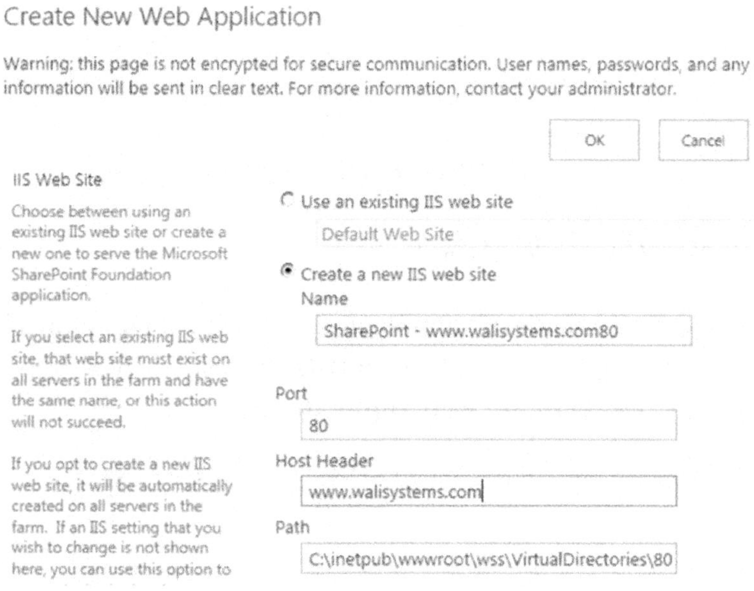

Create New Web Application

Warning: this page is not encrypted for secure communication. User names, passwords, and any information will be sent in clear text. For more information, contact your administrator.

OK
Cancel

IIS Web Site

Choose between using an existing IIS web site or create a new one to serve the Microsoft SharePoint Foundation application.

If you select an existing IIS web site, that web site must exist on all servers in the farm and have the same name, or this action will not succeed.

If you opt to create a new IIS web site, it will be automatically created on all servers in the farm. If an IIS setting that you wish to change is not shown here, you can use this option to

Use an existing IIS web site
Default Web Site
Create a new IIS web site
Name
SharePoint - www.walisystems.com80
Port
80
Host Header
www.walisystems.com
Path
C:\inetpub\wwwroot\wss\VirtualDirectories\80

Figure 7-9: *Create new web application*

6. Enter **80** in the **Port.** Enter **www.walisystems.com** (Your A Record that you created above) in the **Host Header**. Keep all other default values and click **OK**.

7. Next, create a new site collection. On the **Application Management** page, click **Create Site Collections.** From the **Web Application** drop down, select newly created web application.

8. Enter **Title** and **Description** of the site collection.

9. In **Web Site Address** section, in **URL** drop down, select / (front slash) for the root site collection. You can only create one top level site collection per web application. For more site collections, you have to create managed paths. There is a link called **Define Managed Paths** in **Web Site Address** section (on the same page). Also, if you have already used root site collection, **Create Site Collection** page will automatically show you **/sites/** managed path. If you want to use a managed path then you have to enter a site collection name in the text box after the drop down.

10. In the **Template Selection**, keep **2013** selected in the experience version drop down. Keep **Team Site** selected under **Collaboration** tab.

11. Enter **Primary Site Collection Administrator.**

12. Click **Ok** to create site collection.

13. When new web application is created, a default entry is automatically added in the **Alternate Access Mappings**. To see the new entry for your web application, on the **Application Management** page, click **Configure alternate access mappings.** You will see new web application listed and the zone will be **Default.**

14. Now, let's add another URL for the **HR** department. Repeat steps 2 and 3 above to add a new host entry but instead of entering **www**, enter **hr**. **IP** will remain same.

15. Now, you can create a new web application and site collection for the **HR** department just like you did above OR you can use an existing web application. You have already seen how to create new web application and use new URL. Now you will use existing web application that you created above and use **Alternate Access Mappings** to forward new URL to that web application.

16. On the **Application Management** page, click **Configure alternate access mappings** link.

17. Click **Add Internal URLs.**

18. From the **Alternate Access Mapping Collection** drop down, select the newly created web application.

19. Enter new URL (new host entry for the **HR** department, for example, http://hr.walisystems.com) in the **URL protocol, host and port** field.

20. In the **Zone** drop down select **Custom.**

21. Click **Save.**

22. Open new URL (http://hr.walisystems.com) in the browser. Your new site collection will open.

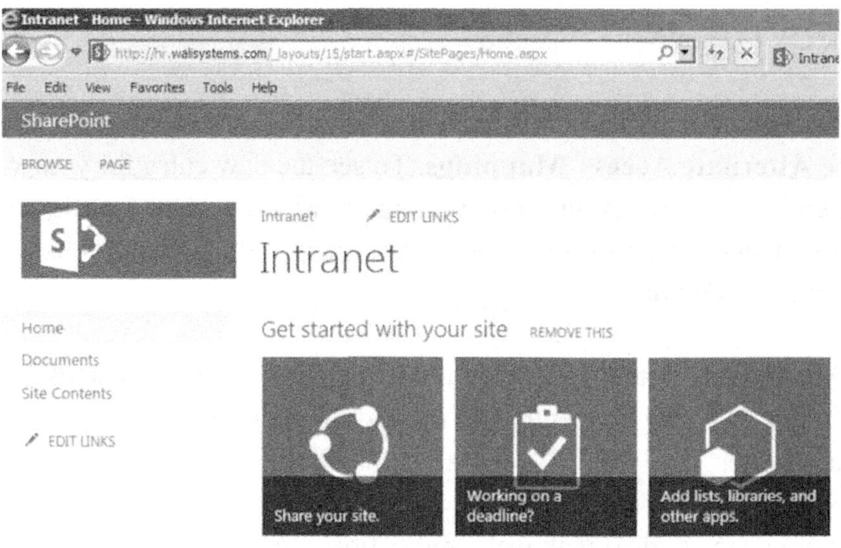

Figure 7-10: *SharePoint site with unique URL*

Summary

In this chapter you learned how to configure alternate access mapping in SharePoint. The chapter discussed creating A record entries and then configuring SharePoint. You also learned how to create a new web application to take advantage of different zones in the web application (AAM) settings.

Chapter 8

Securing SharePoint Sites with Secure Sockets Layer

In this chapter:

- Setup SSL Using Domain Certificate
- Setup SSL Using Self Signed Certificate
- Setup SSL Using Commercial Certificate

Setup SSL in SharePoint 2013 Using Domain Certificate

In this chapter, you will learn how to setup Secure Sockets Layer (SSL) using different types of certificates. Most people get confused as to why SSL is needed in the first place. Especially when you are providing consulting services to a big corporate client and you tell them that to setup an Extranet, SSL is recommended and their first question always is "Why SSL?" The primary reason why SSL is used is to keep sensitive information encrypted. Only the intended recipient should be able to read the information transmitted over the Internet. In the Intranet environment, where you computers are behind a firewall, SSL is not required but in the Extranet and Internet environments, where your information travels outside of your firewall boundaries, SSL is definitely needed especially if the information is sensitive like financial or confidential information. Information when transmitted over the Internet is passed from computer to computer to get to the destination server. Any computer in between you and the server can sniff this information if it is not encrypted with an SSL certificate. SSL certificate encrypts the information and renders it unreadable to everyone except for the server you are sending the information to. This protects the information from hackers.

In addition to encryption, a proper SSL certificate also provides authentication. SSL certificates are issued by a server. If you are purchasing a certificate from a third party company, they usually verify your information before issuing you a certificate. You also get a trust seal from the provider that instills trust in your customers and partners. SSL also protects you from phishing attacks. Hackers are unable to impersonate your site because they don't have a proper SSL certificate.

I can go on and on talking about the benefits and advantages of using SSL with your SharePoint sites but that's not the objective of this chapter. Briefly speaking, if you intend to open up your sites to the public, either through Extranet or through Internet, it is important that you secure your sites with an SSL certificate and that's exactly what this chapter will teach you. In the later chapters, I will teach you how to setup an Extranet site in SharePoint. This chapter is a pre-requisite if you are interested in learning how to setup an Extranet site.

This chapter is divided into three sections. In the first section, you will learn how to setup an SSL using a domain certificate. In the second section, you will learn how to setup an SSL using self-signed certificate and in the final section, you will learn how to setup an SSL using a commercial third party SSL certificate.

Setup SSL Using a Domain Certificate

In this first section, you will first learn how to create a domain certificate. Please note, some or all of these steps may be repeated in the subsequent sections also but that is only for your convenience so that you don't have to go back and forth looking for pertinent information.

To generate a domain certificate, you must have **Active Directory Certificate Services** running.

1. Open **Server Manager** (All Programs > Administrative Tools > Server Manager).

2. Expand **Roles** node and see if certificate services role is installed. If not, then install it first using the steps below.

3. Click **Roles**. Under **Roles Summary** header, you will see **Add Roles** link on the right, click it.

4. Click **Next**.

5. Check **Active Directory Certificate Services** role and click **Next**.

6. Click **Next** again.

7. **Certification Authority** will already be selected. Select the following services:

    a. Certification Authority Web Enrollment

    b. Online Responder

    c. Certificate Enrollment Policy Web Service

8. Click **Next**.

9. Keep **Enterprise** selected and click **Next**.

10. Keep **Root CA** selected and click **Next**.

11. Keep **Create a new private key** selected and click **Next**.

12. Select **RSA#Microsoft Software Key Storage Provider** in the **cryptographic service provider (CSP)**. Change **key character length** from 2048 to 1024 unless this is production environment and you want to use strong keys. By default, **SHA1** is selected in **hash algorithm**, keep it selected and click **Next**.

13. Keep the default values selected and click **Next**. **Common name for CA** is the name that you will see in certification authority while generating domain certificate.

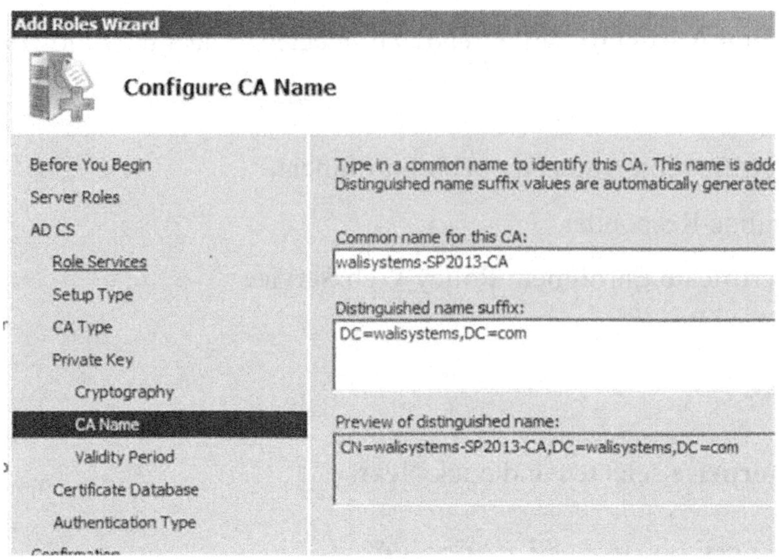

Figure 8-1: *Configure CA Name*

14. Keep default value selected and click **Next**. Default validity period is 5 years.

15. Keep default values selected and click **Next**. These are certificate database and log locations.

16. Keep default option **Windows Integrated Authentication** selected and click **Next**.

17. Click **Install**.

18. Now you have it installed, next step is to create a domain certificate but wait a minute. If your domain controller is on a separate machine then there is one step left. You have to import certificate to the SharePoint machine. The certificate is located in the following folder and has a .crt extension.

    C:\Windows\System32\CertSrv\CertEnroll

    If you ever renamed your server, you will see multiple .crt files. Make sure you pick the one that is current. For example, if your server's FQDN is **walisystems.com** and server name is **SP2013** then the certificate file name will be **sp2013.walisystems.com_walisystems-SP2013-CA**.

Copy the file to the SharePoint machine and import it into **Trusted Root Certification Authorities**.

a. Click Start > Run and type **mmc** and click **OK**. MMC console will open.

b. From **File**, select **Add/Remove Snap-in**.

c. Select **Certificates** from available snap-ins and click **Add >.**

d. Select third option **Computer account** and click **Next.**

e. Choose **Local Computer** and click **Finish**.

f. Click **Ok**.

g. Expand **Certificates (Local Computer)** node.

h. Expand **Trusted Root Certification Authorities** and click **Certificates** folder.

i. Right-click **Certificates** folder and select **All Tasks** then select **Import**.

j. Browse to the certificate (**.crt**) file that you copied from the DNS machine. Click **Next**.

k. Select **Automatically select the certificate store based on the type of certificate** and leave default store selected. Click **Next**.

l. Click **Finish**.

m. You will get **The import was successful** message. Click **Ok**.

19. Now, let's move to the next part which is creating a domain certificate. Open **IIS**.

20. Click on server name and under **Actions** on the right, click **Create Domain Certificate**.

21. Enter a friendly **Common name** for the certificate, for example, your server's FQDN. **Organization** should contain your organization's name or your server name. **Organization Unit** can be an abbreviation of your organization name or machine name. Enter **City**, **State** and select **Country**. Enter full state name, not the abbreviation. Click **Next**.

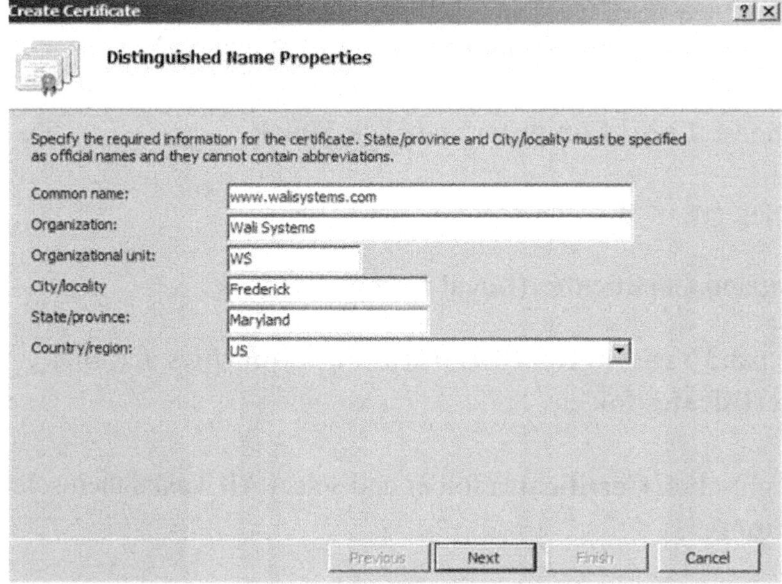

Figure 8-2: *Create Certificate*

22. Click Select button to select **Certificate Authority**. Select the one that you created above. If you are doing this first time then there will be only one authority listed there. Select it and click **OK**. Give a friendly name to the **Online Certificate Authority**, for example, WS_SP2013 and click **Finish**.

That's it. Next you will bind the certificate to your site.

Bind Certificate to Your Site

23. Open **IIS**.

24. Click server name. Expand **Sites** node.

25. Click site name that you will bind to the SSL certificate.

26. On the right, under **Actions**, click **Bindings**.

27. Click **Add**.

28. In **Type**, select **https**.

29. Keep 443 in the **Port**. This is default port used for SSL.

30. In **SSL Certificate**, select the certificate you just created. Look for the common name, for example, WS_SP2013. Click **OK**. That's it.

To test SSL setup, open the site in browser. In the address bar, click the lock sign to check validity of the certificate. If you want to see the certificate, click **View Certificates** link at the bottom of the notification. In case you see error message, click **Certificate Error** (that appears instead of a golden lock). Click View Certificates. Click **Install Certificate** button to install the certificate. Click **Ok** to close the certificate window. Refresh your browser and now you will see a lock.

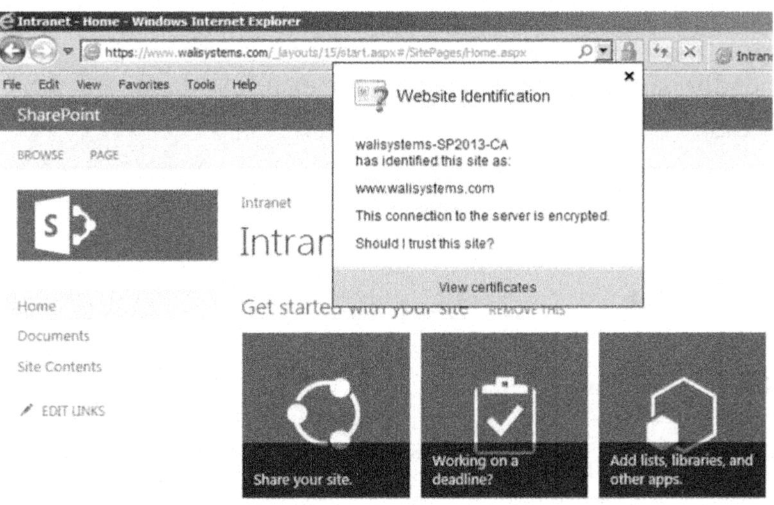

Figure 8-3: *Browse site with SSL enabled*

Setup SSL Using a Self-Signed Certificate

In this section, you will learn how to setup SSL using self-signed certificate. This is useful if you are setting up a development environment or want to test SSL locally. It provides same level of protection as any other commercial certificate and you don't have to pay for it. Why then people use commercial certificates? It's a matter of trust. Basically you are asking your site users to trust you. It may work in some scenarios but might not work if your site is public facing (Internet). Why would people search you, they don't know you. So, you can use self-signed certificate locally for testing but for production you should use a reliable third party certificate issuing authority like Verisign, Thwarte, etc.

Follow the steps below to create a self-signed certificate for your site:

1. Open IIS 7.0.

2. Click on the server name in the navigation tree on the left.

3. On the right side, Under **IIS**, double-click **Server Certificates**.

4. On the extreme right, under **Actions** click **Create Self-Signed Certificate** link.

5. Enter a friendly name for the certificate, for example, I entered WS (for Wali Systems).

6. Double-click the newly created certificate.

7. Click on **Details** tab.

8. Click **Copy to File** button.

9. Certificate export wizard will start. Click **Next**.

10. By default second option **No, do not export the private key** is selected. Keep it selected and click **Next**.

11. Keep the default option **DER encoded binary X.509 (.CER)** selected and click **Next**.

12. Click **Browse** to go to the folder where you want the file to be saved. Enter file name and click **Save**.

13. Click **Next** and then **Finish**. Click **Ok** to close the success message box. Click **OK** to close the **Certificate** window.

The steps that follow are same as those demonstrated in the previous section. Certificates are issued to a computer, user. Or service. Administrators can add certificates to the **Trusted Root Certification Authorities Store** for a local computer or for a domain. Below, we will add certificate to the local computer store.

14. Click Start > Run and type **mmc** and click **OK**. MMC console will open.

15. From **File**, select **Add/Remove Snap-in**.

16. Select **Certificates** from available snap-ins and click **Add >.**

17. Select first option **My user account** and click **Finish.**

18. Click **OK**.

19. Expand **Certificates – Current User** node.

20. Expand **Trusted Root Certification Authorities** and click **Certificates** folder.

21. Right-click **Certificates** folder and select **All Tasks** then select **Import**.

22. Browse to the certificate (**.cer**) file that you saved earlier. Click **Next**.

23. Select **Place all certificates in the following store** and leave default store selected. Click **Next**.

24. Click **Finish**.

25. You will get **The import was successful** message. Click **Ok**.

If you share this server with others, then it's better to import the certificate using local computer account. Follow steps 14 – 25 again but this time in step 17, instead of selecting **My user account**, select **Computer account**. After you have imported certificate into **Trusted Root Certification Authorities**, import it into SharePoint Certificates as well. Expand **SharePoint** node, right-click **Certificates** node and import the certificate.

Manage Trust

26. This step is not required if you have a single server farm but if you are setting it up in a medium or large farm, then you should add certificate to the **Trust Relationships** in central administration site.

27. Open central administration site. Go to **Security** section (Click **Security** under **Central Administration** on the left).

28. In **General Security** section, click **Manage Trust**.

29. In the ribbon, click **New** button.

30. Add a name for this trust relationship.

31. Click **Browse** to import the certificate. This is mandatory regardless of whether you want to provide to or consume trust from the other farm.

32. Leave **Provide Trust Relationship** unchecked unless you want to provide trust to another farm. This is optional.

33. Click **OK**.

You can also add certificate using PowerShell. Open **SharePoint 2013 Management Shell** and run following command:

- $trustcert = new-object system.security.cryptography.x509certificates.x509certificate2("**C:\\ws.cer**")
- New-sptrustedrootauthority –name "**SP Cert**" –certificate $trustcert

** **C:\\ws.cer** is the path to the certificate file. Change it to the path on your machine.
** **SP Cert** is the name that you give to this trust relationship. This is what will appear in the **Manage Trust** interface in SharePoint Central Admin.

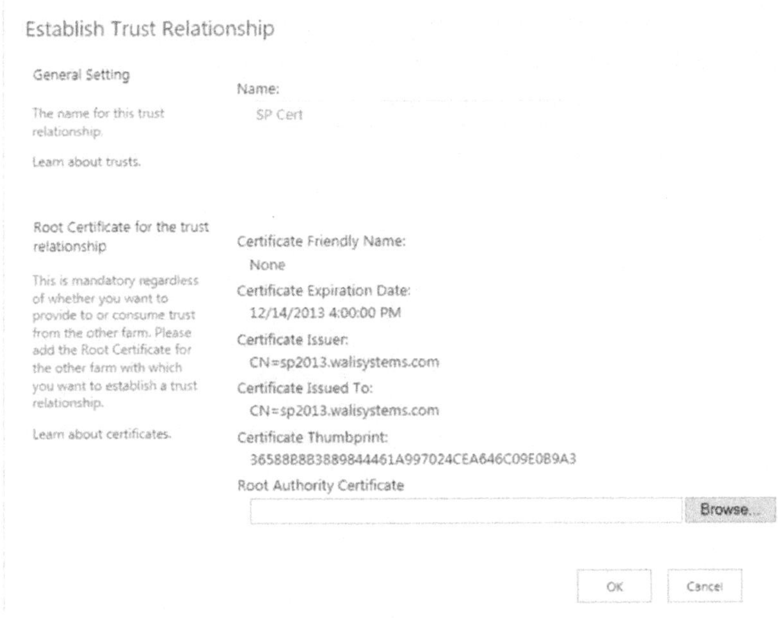

Figure 8-4: *Establish Trust Relationship*

Your site should have correct host headers if you want this certificate to work correctly. For example, if you look at the figure above, you will notice that the certificate was issued to **sp2013.walisystems.com** so if your site does not have this host header, you will get an error. For example, when you try to open your site in the browser, you get this error:

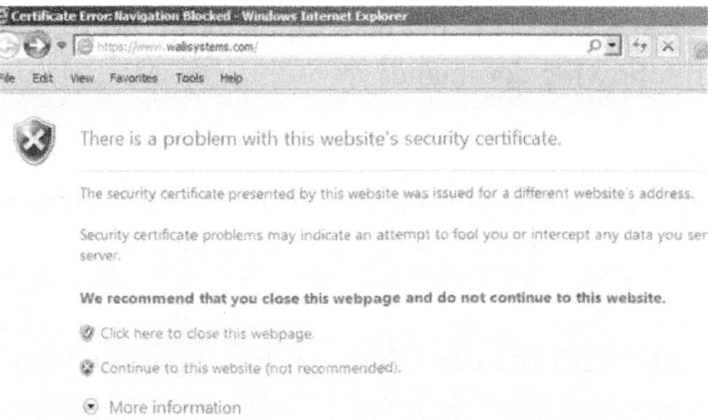

Figure 8-5: *There is a problem with this website's security certificate*

Again, notice that I tried to open **https://www.walisystems.com** but the certificate was issued to **sp2013.walisystems.com**. If you click **Continue to this website**, site will open but you will still get an error message.

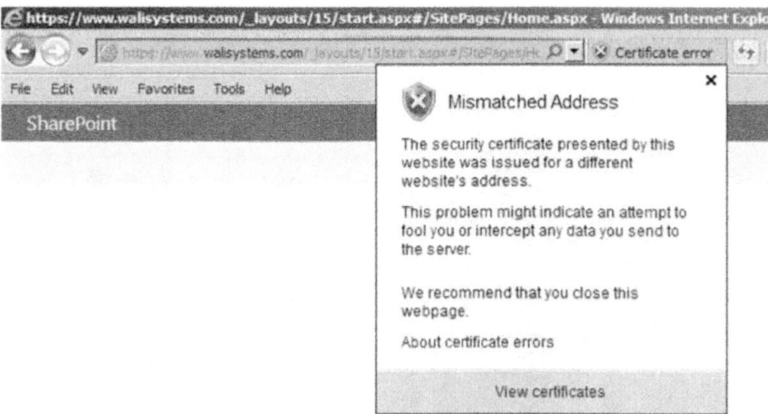

Figure 8-6: *Mismatched Address*

Error: *The security certificate presented by this website was issued for a different website's address. This problem might indicate an attempt to fool you or intercept any data you send to the server.*

To resolve the issue, create a new web application or extend an existing one. If you want to add https to your main site at port 80, then create a new web application with new host headers. Here are the steps:

Create New Web Application for SSL

34. Go to central administration. Click **Manage web applications**.

35. Click **New** button in the ribbon.

36. Select **Create a new IIS web site**. Change **Port** to **80**.

37. In the **Host Header**, enter the URL that you want to use for this web application. For example, I wanted to use "sp2013.walisystems.com" because that was the URL for which the certificate was issued therefore I entered "sp2013.walisystems.com". Note: Do not add **HTTP** in the URL.

38. In **Security Configuration** section, select **Yes** in **Use Secure Sockets Layer (SSL)**.

39. Keep all other default options selected and click **OK**.

40. After web application is created, create a site collection at the root level.

Change Alternate Access Mappings

41. In Central Administration Site, go to **Application Management** section and click **Configure alternate access mappings**.

42. Change site collection in the drop down. Select the one that you just created. Click **Add Internal URLs**.

43. Enter complete URL that starts with **HTTPS**. For example, **https://sp2013.walisystems.com**.

44. Change **Zone** to **Custom** or **Extranet**.

45. Click **Save**.

Bind Certificate to Your Site

46. Finally, bind certificate to your site. Open **IIS**.

47. Click server name. Expand **Sites** node.

48. Click site name that you will bind to the SSL certificate.

49. On the right, under **Actions**, click **Bindings**.

50. Click **Add**.

51. In **Type**, select **https**.

52. Keep 443 in the **Port**. This is default port used for SSL.

53. In **SSL Certificate**, select the certificate you just installed. Click **OK**. That's it.

To test SSL setup, open the site in browser. In the address bar, click the lock sign to check validity of the certificate. If you want to see the certificate, click **View Certificates** link at the bottom of the notification.

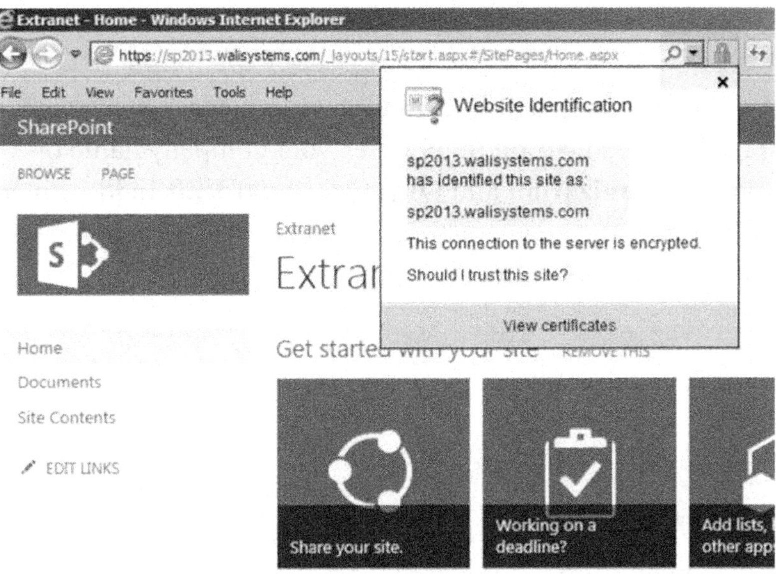

Figure 8-7: *Browse Extranet site with SSL enabled*

Setup SSL Using Commercial Certificates

In this section, you will learn how to setup SSL in SharePoint 2013 using commercial certificates.

So, let's start. I will now show you how to use Verisign certificate. Verisign is one the most popular companies that issue SSL certificates. We will sign up for a trial version.

1. Before you sign up on Verisign site, we first need to create a certificate request. This will be needed when you sign up at Verisign.

2. Open IIS 7.0 (Start > Administrative Tools > Internet Information Services (IIS) Manager).

3. Click on the server name.

4. In **IIS** section, double-click **Server Certificates**.

5. On the right side, under **Actions**, click **Create Certificate Request…** link.

6. **Request Certificate** form will open. Fill out the fields. Enter your site name or URL in the Common **Name** field. Enter your company name or abbreviation in **Organization** and **Organizational Unit** fields. Enter **City** and **State**. Enter full state name, abbreviation is not accepted. Select **Country/region** and click **Next.**

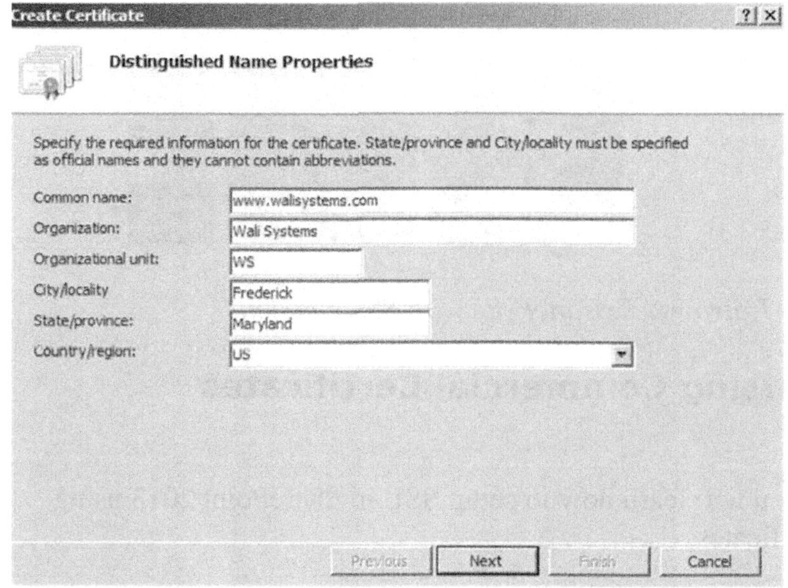

Figure 8-8: *Request Certificate*

7. Keep the default values selected. **Cryptographic service provider** should have **Microsoft RSA SChannel Cryptographic Provider** selected. If it's a test or development environment, you can keep **Bit length** set to 1024. If it's a production environment and you are using a purchased SSL key, then select **Bit length** according to your needs. What kind of security is needed depends on what kind of site you have created and what kind of content it has. For confidential content or for government sites, you may want to select at least 2048 in **Bit length.** Remember, the greater the bit length, the stronger the security. However, a greater bit length may decrease performance.

8. Browse to a folder that will store the request and give a name to the file, for example, sslrequest. Click **Finish.**

9. Go to Verisign site and sign up for a trial version. Here is the direct link for the signup page:

   https://trustcenter.websecurity.symantec.com/process/retail/trial_product_selector;jsessionid=5FD7750365927477161 7C69ABB11A28D?uid=f0e5b79664a05e9b52b4844f12670f91&locale=VRSN_US

   Enter technical contact details and click **Continue.** On the next screen, you will be asked to enter CSR. Open the request file that you had created, copy it and paste it into the box on the site. Once you have signed up, you will get an email with the key.

10. There will be three links in the email. Click the first link to download and install the Test Root CA Certificate. On the download page, there are different browsers listed. Select the browser that you will use for your site testing. Remember if you know you audience will use different browsers then you need to perform this step for every browser that your audience will be using. Steps for Internet Explorer are listed next.

11. Click the link **Download Secure Site Trial Root Certificate** link. From the box, copy the certificate and save it in a text file with a .cer extension.

12. Open Internet Explorer.

13. Go to Tools > Internet Options > Content > Certificates.

14. Click **Import….** A wizard will open. Click Next.

15. Browse to the location of the recently stored .cer file (step 11 above).

16. Select the certificate and click **Open.**

17. Click **Next.**

18. Select **Automatically select the certificate store based on the type of the certificate.** Click **Ok.**

19. Click **Next** then **Finish.**

20. When prompted and asked if you wish to add the following certificate to the root store, click **Yes.**

21. Second step listed in the email confuses many people. Basically this step is not required for latest IIS server. Users using IIS 5.0 or Higher servers do not need to download the intermediate CA as it is included with the SSL certificate upon issuance if they selected in the purchase as server vendor: Microsoft IIS 5.0 or higher. If you are not sure about the selection you made when requesting SSL certificate, go ahead and install it. It will not harm anything. To install it, perform same steps as listed above (Steps 11 – 19). Be sure to click second link in your email instead of first when performing step 11.

22. Just like the second step, third step in the email is equally confusing. Don't follow those steps, follow the ones listed below to install the certificate without hassles.

23. Copy the certificate from your email. It will be at the bottom of the email. Be careful when copying. Copy whole text including -----BEGIN CERTIFICATE----- and -----END CERTIFICATE----- and paste it into Notepad. There should be no white spaces or extra line breaks. Save the file with **.p7b** extension. You can name the file as sslcert.p7b or whatever you prefer.

24. Click Start > Run and type **mmc** and click **OK**. MMC console will open.

25. From **File**, select **Add/Remove Snap-in**.

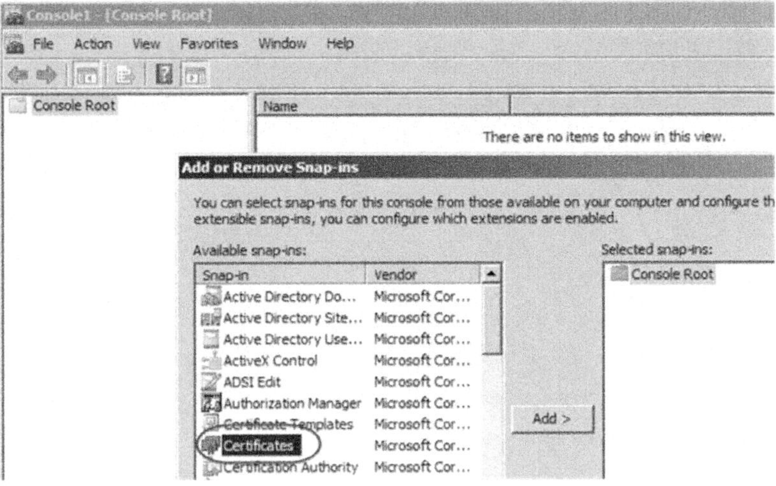

Figure 8-9: *Add Certificates snap-in*

26. Select **Certificates** from available snap-ins and click **Add >.**

27. Select first option **My user account** and click **Finish.**

28. Click **OK**.

29. Expand **Certificates – Current User** node.

30. Expand the **Trusted Root Certification Authorities** folder on the left and select the **Certificates** subfolder.

31. Locate the following certificate:

    a. Issued To: **VeriSign Class 3 Public Primary Certification Authority - G5**

32. **Right-click the certificate and select Properties.**

33. In the **Certificate purposes** section, select **Disable all purposes for this certificate**. This is a pre-installed certificate and must be disabled before using new certificate from SSL.

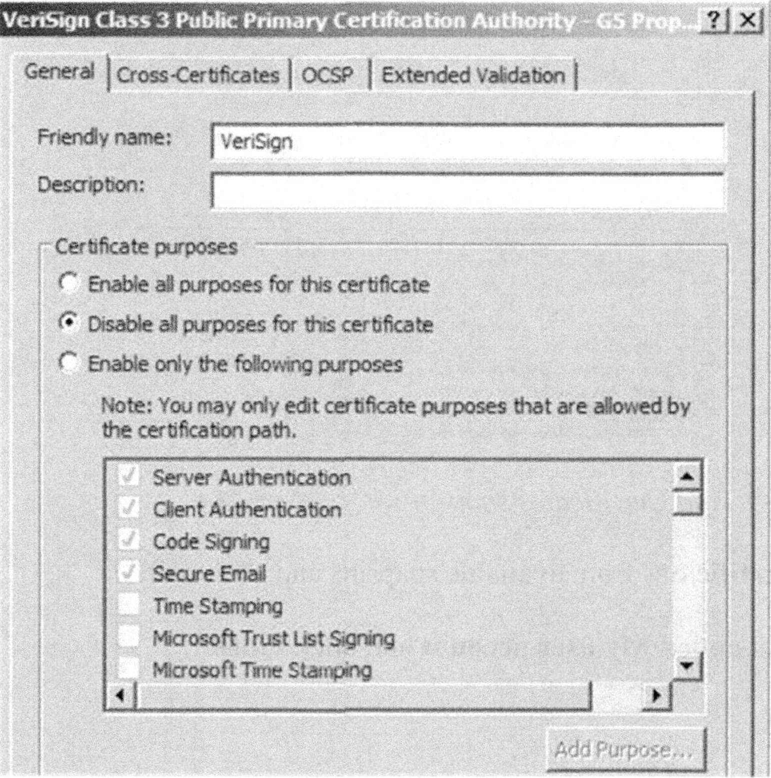

Figure 8-10: *VeriSign SSL Setup*

34. Click **Apply** then **OK**.

Install Certificate

35. Finally, install the certificate. There are two ways to do it. One is through the IIS and the other is the MMC certificates console. I prefer the second way because you have to come to the MMC certificates console anyway to fix a problem if you opt to install the certificate through IIS. If you install the certificate through IIS, at the time of binding the certificate to your site, certificate will not show up in IIS and then you will have to come to the MMC certificates console to perform an additional step. So it's better to go with the MMC route from the beginning. There is another problem with the IIS method. You get the following error when you install the certificate through IIS.

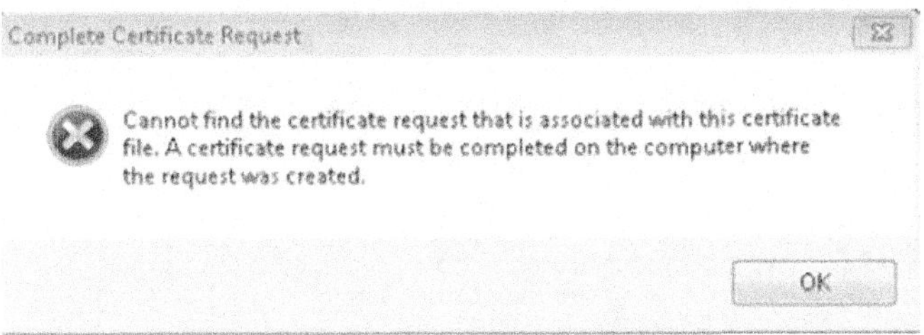

Figure 8-11: *Complete Certificate Request*

Error Text: *Cannot find the certificate request that is associated with this certificate file. A certificate request must be completed on the computer where the request was created.*

36. Open MMC console (Start > Run > Type MMC > Click OK).

37. Certificates console will still be available because you added it in step 24 but if for some reason you had to restart your machine or log out of it then you will have to add the console again. Follow steps 24 – 28 to add **Certificates** console. Expand **Certificates – Current User** node on the left.

38. Expand **Trusted Root Certification Authorities** and click **Certificates** folder.

39. Right-click **Certificates** folder and select **All Tasks** then select **Import**.

40. Browse to the certificate (**.p7b**) file. Click **Next**.

41. Select **Place all certificates in the following store** and leave default store selected. Click **Next**.

42. Click **Finish**.

43. You will get **The import was successful** message. Click **Ok**. You may also get following security warning.

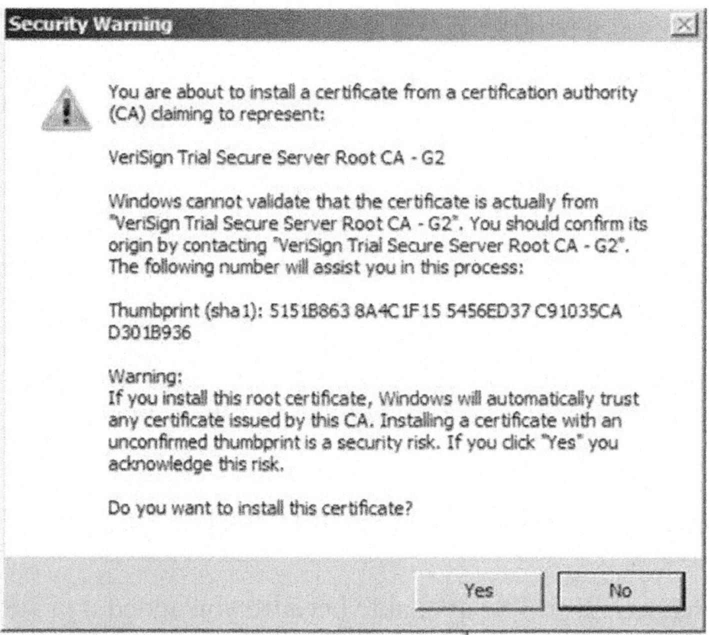

Figure 8-12: *Security warning – You are about to install a certificate ...*

Error Description: *You are about to install a certificate from a certification authority (CA) claiming to represent: Verisign Trial Secure Server Root CA – G2.*

If you get this security warning, click **Yes** to install the certificate.

44. With MMC Certificates console still open, expand **Personal** folder on the left and right-click **Certificates** subfolder. Select **All Tasks** then select **Import**.

45. Browse to certificate file (.p7b) and Click **Next**.

46. Keep second option **Place all certificates in the following store** selected and keep **Personal** certificate selected as the default option. Click **Next**.

47. Click **Finish**. You will get **The import was successful** message. Click **Ok**.

Bind Certificate to your site

48. Finally, bind certificate to your site. Open **IIS**.

49. Click server name. Expand **Sites** node.

50. Click site name that you will bind to the SSL certificate.
51. On the right, under **Actions**, click **Bindings**.

52. Click **Add**.

53. In **Type**, select **https**.

54. Keep 443 in the **Port**. This is default port used for SSL.

55. In **SSL Certificate**, select the certificate you just installed. Please note that if you don't see new certificate in this drop down, then you probably missed steps 44 – 47 above. Click **OK**. That's it.

Alternate Access Mappings

56. Assuming you have a web application setup to work with SSL, configure Alternate Access Mappings to use site with SSL. Open **Central Admin Site** and click **Application Management**.

57. Under **Web Applications**, click **Configure alternate access mappings**.

58. You will notice you already have default site listed in the **Default** zone. To add new URL in the **Intranet** zone, click **Add Internet URLs**.

59. From the **Alternate Access Mapping Collection** drop down, select correct application that you want to use for the AAM setting and then add URL in the text box labeled **URL protocol, host and port**, for example, **https://www.walisystems.com**. From the **Zone** dropdown, select **Intranet**.

60. Click **Save**.

Now open site with https to test that everything works fine.

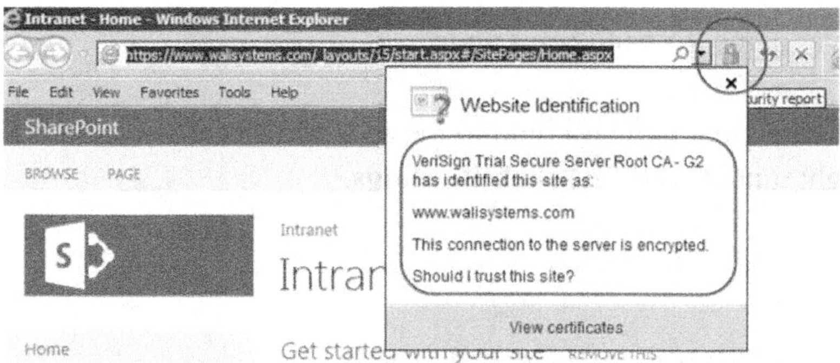

Figure 8-13: *Browse SharePoint site with SSL enabled*

In the address bar, click the lock sign to check validity of the certificate. If you want to see the certificate, click **View Certificates** link at the bottom of the notification.

Summary

This chapter discussed in detail the options for securing SharePoint sites. You learned how to acquire certificates using different means and setting them up for usage in SharePoint. You also learned how to configure DNS and IIS in order to make it work perfectly.

Chapter 9

Setup SharePoint 2013 Extranet

In this chapter:

- Learn How To Setup External Database
- Learn To Use MembershipSeeder Tool
- Learn To Use FBA Configuration Manager
- Learn To Configure Config Files Manually

We have covered all important topics that lay ground work for creating Extranet sites. Here is what is needed for creating Extranets:

1. Web application and site collection setup
2. SSL setup – This is optional but highly recommended if you are allowing data exchange with external entities.
3. Binding SharePoint site with SSL certificate
4. Setup of users database
5. Central admin and web.config configurations
6. Testing

Remember, most of the topics mentioned above are optional. For example, you may not want to have a unique URL for your extranet site. You may be ok with the machine name in the URL but in case, if you want to have a unique URL for your site, read chapters on AAM and Alternate URLs.

There are different options for setting up Extranets. You can use LDAP, you can use custom database, or you can use OpenID. In this chapter, you will learn to use custom database. By default, authentication type in SharePoint 2013 is claims-based. To access custom database or data source like LDAP (Lightweight Directory Access Protocol), forms-based authentication has to be configured in web application settings. Forms-based authentication is an identity management system that is based on ASP.NET membership and role provider authentication.

SharePoint 2013 supports a variety of authentication methods and authentication providers for the following authentication types:

- Windows authentication
- Forms-based authentication
- SAML token-based authentication

Advantage of using claims-based authentication is that you can setup multiple authentication methods in one zone but there are two restrictions that apply:

- You can implement only one instance of forms-based authentication on a zone.
- Central administration allows you to use both an Integrated Windows method and Basic at the same time.

Setup database

To use SQL Membership Provider, first you need to install **ASPNETDB** database. ASP.NET includes a tool for installing the SQL Server database used by the **SqlMembershipProvider** and **SqlRoleProvider**. The **Aspnet_regsql.exe** tool is located in the ***drive:\Windows\Microsoft.Net\Framework64\v4.0.30319***.

If you try to create web application that uses Forms-based authentication using SQL Membership Provider database, you get error message: **Sorry, Something went wrong** and it gives you a correlation ID which when you check in the logs shows several errors, for example, the following:

Microsoft.SharePoint.Administration.Claims.SPClaimEncodingManager.DecodeClaimFromFormsSuffix(String encodedValue) at Microsoft.SharePoint.Administration.Claims.SPClaimProviderManager.GetProviderUserKey(IClaimsIdentity claimsIdentity, String encodedIdentityClaimSuffix) at Microsoft.SharePoint.Administration.Claims.SPClaimProviderManager.GetProviderUserKey(String encodedIdentityClaimSuffix) at Microsoft.SharePoint.Utilities.SPUtility.GetFullUserKeyFromLoginName(String loginName) at Microsoft.SharePoint.ApplicationRuntime.SPHeaderManager.AddIsapiHeaders(IHttpContext context, String encodedUrl, NameValueCollection headers)

The error occurs because Membership provider and Role manager have not been configured yet. We can configure web.config first and then create the application but let's first install the database.

1. Double-click **Aspnet_regsql.exe** to start the wizard.

2. Click **Next**.

3. Choose first option **Configure SQL Server for application services**. You can use same wizard to remove application services information from an existing database. Click **Next**.

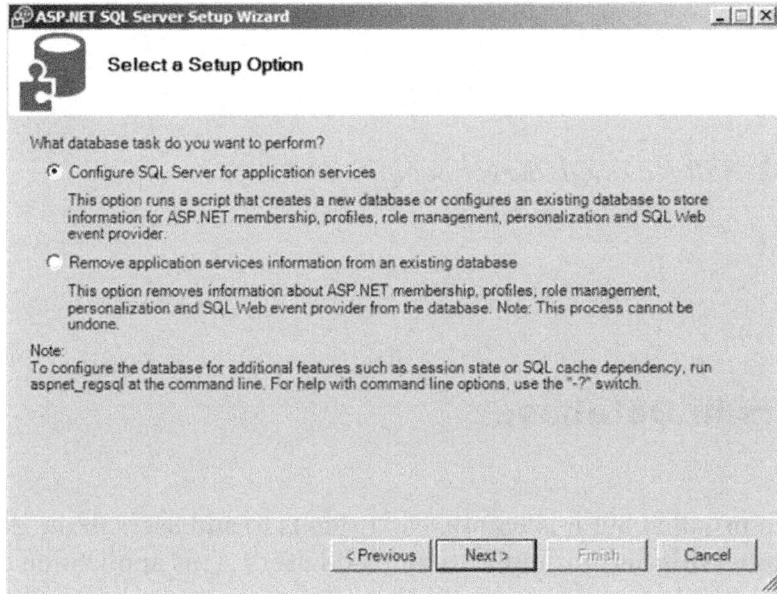

Figure 9-1: *Select a setup option*

4. Enter database server name. Default method for accessing database is **Windows authentication**. If you have set up SQL server user ID and password, you can choose **SQL Server authentication**. Click **Next**.

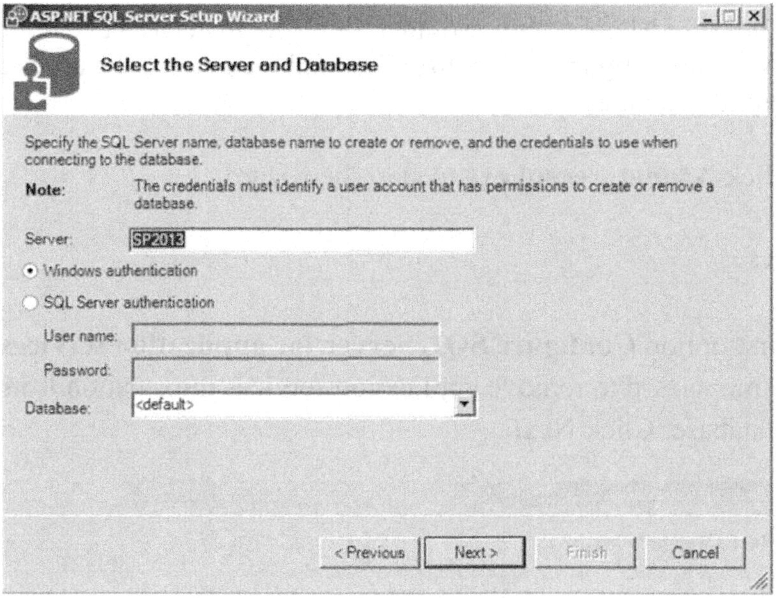

Figure 9-2: *ASP.NET SQL Server Setup*

5. Click **Next**.

6. Click **Finish**.

Adding Users in Database

Database has been installed but it is empty. Next step is to add users in the database. You can use **MembershipSeeder** application to add users. This application can be downloaded from codeplex:

http://cks.codeplex.com/releases/view/7450

Second option is to build your own .NET application that will provide interface to add data in the database.

Third option, the one I used, is to add data manually in the database. This is ok if you want to add users for testing may be once or twice but in production environment where adding and removing users is an ongoing task, you need to have an interface to do this easily. Below, I show you how to add some test data so that we can continue with the application setup but before adding test data, we must give permissions to the accounts that will access this database.

7. Open **Microsoft SQL Server Management Studio.**

8. First we need to add account that we will use in new web application to the newly created database. You can use admin account in the new web application but standard practice is to use special service account for SharePoint applications. I have named this service account "spservice" and I will add it to the database permissions. Also you need to check what account **STS (SecurityTokenService)** uses. Open IIS manager.

Figure 9-3: *Application pools*

Expand server name node, click **Application Pools**. Check **Identity** column again **SecurityTokenServiceApplicationPool**. This account (WALISYSTEMS\Administrator) must be added in the database permissions.

9. There are two ways to accomplish this. You can add these accounts directly in the **ASPNETDB** database permissions or you can add them to the **sysadmin** role, that will automatically give them permissions on all databases. To add these accounts in **sysadmin** role, do the following:

    a. In management studio, expand **Security** node.

    b. Expand **Server Roles**.

    c. Double-click **sysadmin**.

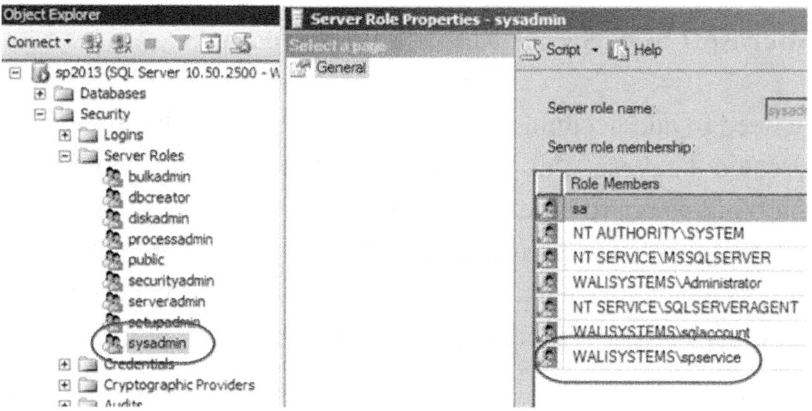

Figure 9-4: *Server role properties*

d. If you don't see accounts listed there then you need to add them first in **Logins**. Right-click **Logins** node and select **New Login**. Enter **Login name** and click Search. Click **OK** if login name is found otherwise click **Check Names** to find the login and then click **OK**.

e. Click **OK**.

f. Login has been added. Now it's time to add it to the server roles. Expand **Server Roles** node, then double-click **sysadmin** node.

g. Click **Add**.

h. Enter Login name and click **Check Names**. Click **Ok** to add login to the role.

Repeat these steps to add the second login. One was **Walisystems\spservice** account that was especially for SharePoint. Second was **WaliSystems\Administrator** account that STS application pool uses.

To add these accounts directly to the database permissions (instead of adding them in **sysadmin** role), do the following:

a. Expand **aspnetdb** database node.

b. Expand **Security** node.

c. Right-click **Users** and select **New User**.

d. Enter **User name**. Enter same in the **Login name** and click ellipsis. Enter login name, click **Check Names** then click **Ok**.

e. Check following options in both **Owned Schemas** and **Database role membership** boxes:

- Aspnet_Membership_FullAccess
- Aspnet_Personalization_FullAccess
- Aspnet_Profile_FullAccess
- Aspnet_Roles_FullAccess
- Aspnet_WebEvent_FullAccess

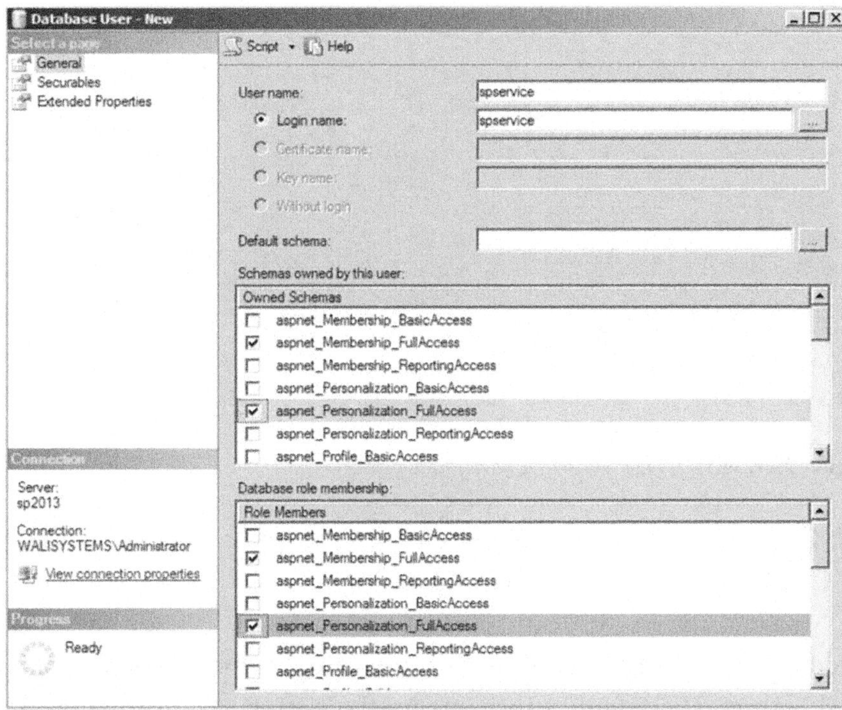

Figure 9-5: *Database user*

10. Now that accounts have been added, next step is to add data. I will show you two methods for adding data, one using the **MembershipSeeder** tool and

second adding users manually. First, I will show you how to use **MembershipSeeder** tool.

Add Users Using MembershipSeeder Tool

11. Download **MembershipSeeder.zip** from following location:

    http://cks.codeplex.com/releases/view/7450

12. Unzip package and go to **MembershipSeeder > Bin > Release** folder and run **MembershipSeeder.exe**.

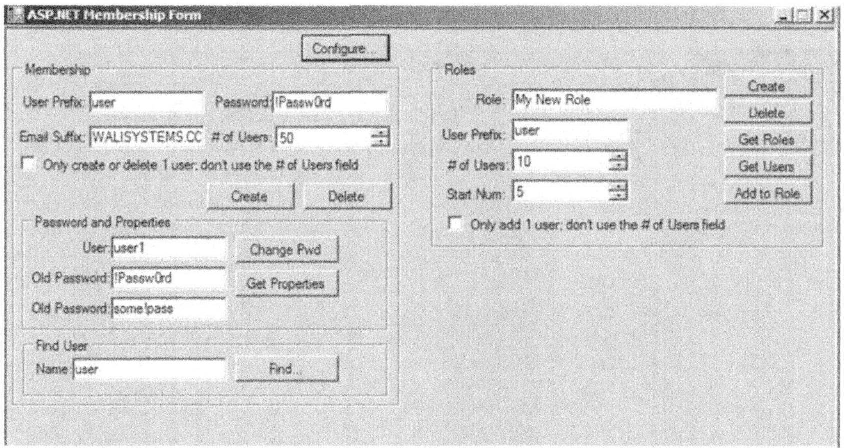

Figure 9-6: *ASP.NET Membership Form*

13. Click **Configure** to enter database name. Enter your database name. Click **OK**. Close the application and re-open.

14. To add users, click **Create** in the **Membership** section. You can change the default values. By default, username prefix is **user**. All usernames will have this prefix. Password can also be changed. Enter your domain name in the **Email Suffix**. By default, application will show your server's FQDN. Finally enter number of users that you want to create. Default number is 50. If you want to create only one user, check **Only create or delete 1 user; don't use the # of Users** field. This will only create one user. If you chose

more than one user, then application will create all users with **user** prefix. For example, you created two users, usernames will be user1 and user2. You can also delete users by clicking the **Delete** button.

Results | Messages

	ApplicationId	UserId	UserName	LoweredUserName	MobileAlias	IsAnonymous
1	B0181FCE-0E0B-425A-91A8-6CB0A4FDAE77	4E2F1DD4-4BAC-4F00-A285-E50589C3AF37	user	user	NULL	0
2	B0181FCE-0E0B-425A-91A8-6CB0A4FDAE77	1C4BDEA3-37D1-4B0B-AAC0-240E27D9F0E8	user1	user1	NULL	0
3	B0181FCE-0E0B-425A-91A8-6CB0A4FDAE77	10FC5511-1258-428B-8AF4-208D6F488F51	user2	user2	NULL	0
4	9F369FB1-9480-4D8A-BB44-7838A3C2AFA8	9F369FB1-9480-4D8A-BB44-7838A3C2AFAC	saf	saf	saf	0

Figure 9-7: *aspnet_Users table*

15. To add roles, enter **Role** name. You can change **User Prefix, # of Users, and Start Num**. Click **Create** to create the role. I needed to add an admin role, therefore, I entered **Admin** in the **Role** field. I changed **User Prefix** to **Admin** and **# of Users** to **1**. **Start Num** should be **1** if this is first admin you are adding.

Results | Messages

	ApplicationId	RoleId	RoleName	LoweredRoleName	Description
1	B0181FCE-0E0B-425A-91A8-6CB0A4FDAE77	889E077F-E146-4614-988B-3601912AA5F8	Admin	admin	NULL

Figure 9-8: *aspnet_Roles table*

16. To change password, use **Password and Properties** section. To find a user, use **Find User** section. To see all roles, click **Get Roles** button. To see all users, click **Get Users** button. To add user to a role, click **Add to Role** button.

Add Users Manually

17. If you have access to the tool, then why would you add users manually? This process will help you understand the database structure and you will learn what data needs to go to what tables. In SQL Server Management Studio, click **New Query** button.

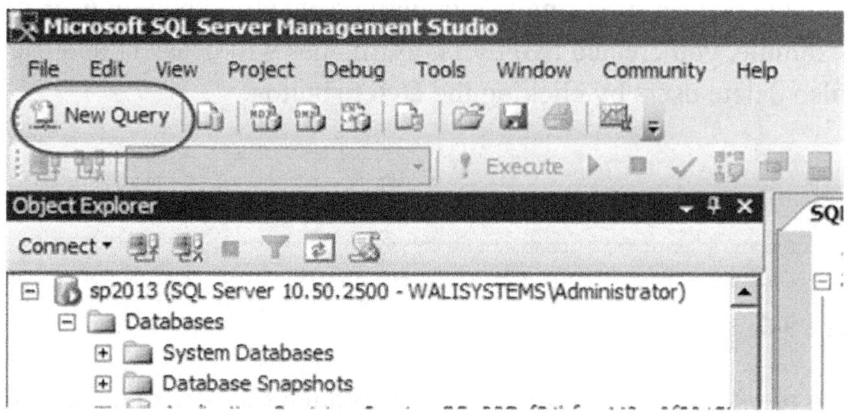

Figure 9-9: *New Query button*

18. You need three GUIDs, one for **ApplicationId**, one for **RoleId** and one for **UserId**. We will add just one user for now. For every new application, role and user you need a new GUID. That's where data tool comes in handy (discussed above). Now, we will generate GUIDs manually in the SQL script window. In the script window, write following command and click **Execute** button. Run this command thrice to create three GUIDs and note them down, you will need them when adding data to tables.

    Select newid()

19. Right-click **aspnet_Applications** table and select **Edit Top 200 Rows**. Add following data in new row:

Column	Value
ApplicationName	Fba
LoweredApplicationName	fba
ApplicationId	9f369fb1-9480-4d8a-bb44-7838a3c2afab
Description	fba

Table 9-1: *Applications Table*

20. Add following data to the **aspnet_Roles** table.

Column	**Value**
ApplicationId	9f369fb1-9480-4d8a-bb44-7838a3c2afab
RoleId	9f369fb1-9480-4d8a-bb44-7838a3c2afaa
RoleName	Admin
LoweredRoleName	admin
Description	Admin Role

Table 9-2: *Roles Table*

21. Add following data to the **aspnet_Users** table.

Column	**Value**
ApplicationId	9F369FB1-9480-4D8A-BB44-7838A3C2AFAB
UserId	9F369FB1-9480-4D8A-BB44-7838A3C2AFAC
UserName	saif
LoweredUserName	Saif
MobileAlias	NULL
IsAnonymous	0
LastActivityDate	2013-12-23

Table 9-3: *Users Table*

22. Add following data to the **aspnet_UsersInRoles** table.

Column	**Value**
UserId	9F369FB1-9480-4D8A-BB44-7838A3C2AFAC
RoleId	9F369FB1-9480-4D8A-BB44-7838A3C2AFAA

Table 9-4: *UserInRoles Table*

23. Add following data to the **aspnet_Membership** table.

Column	Value
ApplicationId	9F369FB1-9480-4D8A-BB44-7838A3C2AFAB
UserId	9F369FB1-9480-4D8A-BB44-7838A3C2AFAC
Password	Saif
PasswordSalt	wRlFK+hhgCjVkJXSxHtaYA==
MobilePIN	NULL
Email	SAIF@WALISYTEMS.COM
LoweredEmail	saif@walisystems.com
PasswordQuestion	NULL
PasswordAnswer	NULL
IsApproved	1
IsLockedOut	0
CreateDate	2013-12-23
LastLoginDate	2013-12-23
LastPasswordChangedDate	2013-12-23
LastLockoutDate	1900-01-01
FailedPasswordAtemptCount	0
FailedPasswordAttemptWindowStart	1900-01-01
FailedPasswordAnswerAttemptCount	0
FailedPasswordAnswerAttemptWindowStart	1900-01-01
Comment	NULL

Table 9-5: *Membership Table*

That is it! Data has been added. Next, you will create new web application.

Create a new web application that uses forms-based authentication

24. Login to central administration site.

25. In **Application Management** section, click **Manage web applications** link.

26. Click **New** button in the ribbon.

27. Enter site name, for example, I named it "Development" because I wanted to collaborate with external developers.

28. Enter **80** in the Port.

29. Enter URL in the **Host Header**. For example, my server's FQDN is **walisystems.com** and I wanted the extranet site to have **dev.walisystems.com** URL therefore I entered **dev.walisystems.com** in the **Host Header**.

30. If you are sure that you are going to setup SSL for this web application, then set **Use Secure Sockets Layer (SSL)** to **Yes** otherwise keep it **No** and you can change it later if needed.

31. Keep **Enable Windows Authentication** checked. **Integrated Windows Authentication** should also be checked. In the drop down, select **NTLM**. Remember, if you want the site to be used purely for external users then you don't need to enable Windows authentication. Windows authentication will allow internal users to access the site. If it's a collaboration site and internal users will collaborate with external users, then Windows authentication should be enabled.

32. Check **Enable Forms Based Authentication (FBA)**.

33. In the **ASP.NET Membership provider name**, enter membership name that you will use in the configuration. It can be anything. For example, membeshipprovider or simply External, etc. I used **External**.

34. In the ASP.NET Role manager name, enter role manager name that you will use in the configuration, for example, roleprovider. I used **RoleManager**.

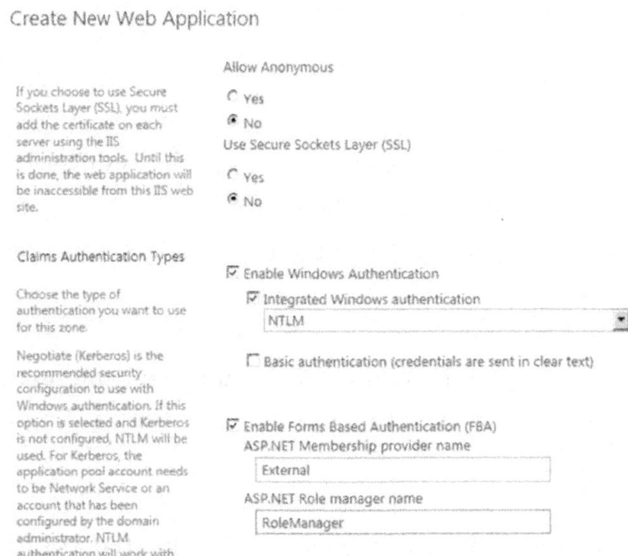

Figure 9-10: *Create new web application*

In the application pool, I used special account that I had created for SharePoint. This is called SharePoint service account. In the test environment, you can use your admin account.

Figure 9-11: *Application pool*

35. Keep other default options selected and click **OK**.

Web application has been created. If you want to use SSL with your application, you must setup SSL and configure Alternate Access Mappings as described in the previous chapters.

Next step is configuring web.config files.

Configure web.config

Web.config for the extranet site and STS site has to be configured. Take backup of both config files before you make any changes. If something goes wrong you can revert back to the original file. web.config for the STS application is located in the following folder:

drive:\Program Files\Common Files\Microsoft Shared\Web Server Extensions\15\WebServices\SecurityToken

Another way of finding correct web.config is to use IIS. If you are not sure which web.config to select, go to Internet Information Server (IIS) Manager, and open the **SharePoint web services** site.

In the console, right-click **SecurityTokenServiceApplication**, and then click **Explore**.

In the folder window, double-click the web.config file.

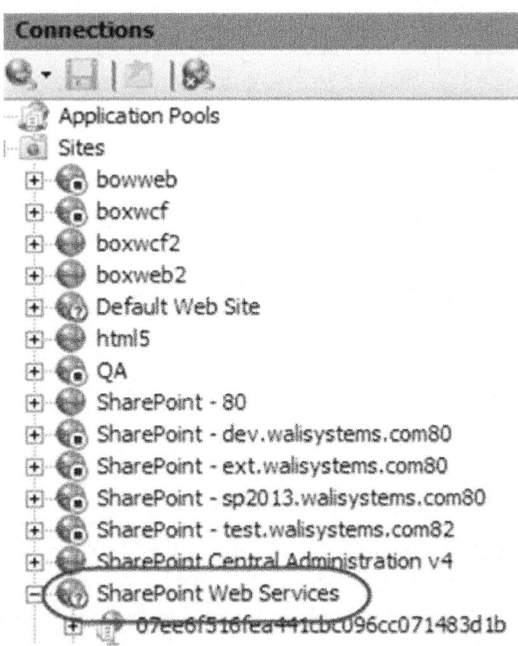

Figure 9-12: *SharePoint Web Services Site*

web.config for the extranet application is located in the following folder:
drive:\inetpub\wwwroot\wss\VirtualDirectories\dev.walisystems.com80

dev.walisystems.com80 is the name of my extranet site. Yours will be different depending on what name you gave to the application.

There are two ways to edit these files. I will explain both below. One is to use FBA Configuration Manager for SharePoint 2013 that can be downloaded from the following link:

http://blogs.technet.com/b/speschka/archive/2012/07/28/fba-configuration-manager-for-sharepoint-2013.aspx

Other way is to edit the files manually. I will explain second option (manually editing) first.

24. Open STS web.config in the editor of your choice.

25. Locate **</system.net>** tag in the file. You need to insert following snippet after this tag. Remember the following snippet will appear between **</system.net>** and **</configuration>** tags.

```
<system.web>
  <membership>
   <providers>
    <add connectionStringName="aspnetdb" applicationName="fba"
name="External"
type="System.Web.Security.SqlMembershipProvider,
System.Web,&#xD;&#xA; Version=2.0.3600.0, Culture=neutral,
PublicKeyToken=b03f5f7f11d50a3a" />
   </providers>
  </membership>
  <roleManager>
   <providers>
    <add connectionStringName="aspnetdb" applicationName="fba"
name="RoleManager" type="System.Web.Security.SqlRoleProvider,
System.Web,&#xD;&#xA; Version=2.0.3600.0, Culture=neutral,
PublicKeyToken=b03f5f7f11d50a3a" />
   </providers>
```

```
    </roleManager>
  </system.web>
  <connectionStrings>
 <add name="aspnetdb" connectionString="Data Source=sp2013;Initial
 Catalog=aspnetdb;Integrated Security=SSPI" />
  </connectionStrings>
```

Listing 9-1: *web.config file*

connectionStringName can be anything, whatever you use in the **connectionString**. Note that we used "aspnetdb" in the **<connectionStrings> </connectionStrings>**, therefore, we used that name in the **<providers> </providers>** tag.

applicationName is the name that you will give to your application when defining data in the database. It can be anything. I preferred "fba". You can name it something else.

External is **MembershipProvider** name that we will use when setting up web application. Similarly, **RoleManager** is **RoleProvider** name that we will use when setting up web application.

Here is the screenshot showing inserted snippet:

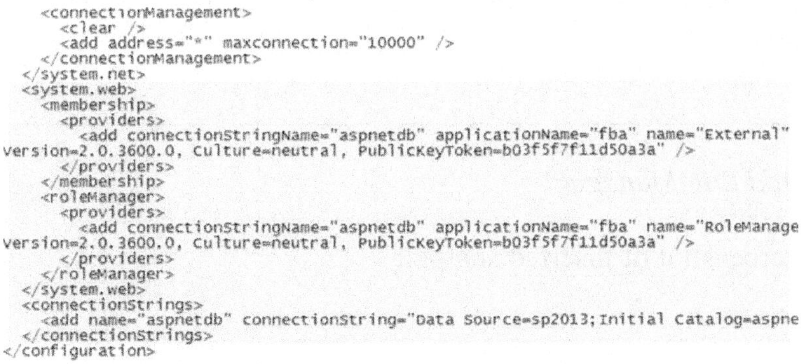

```
    <connectionManagement>
      <clear />
      <add address="*" maxconnection="10000" />
    </connectionManagement>
  </system.net>
  <system.web>
    <membership>
      <providers>
        <add connectionStringName="aspnetdb" applicationName="fba" name="External"
Version=2.0.3600.0, Culture=neutral, PublicKeyToken=b03f5f7f11d50a3a" />
      </providers>
    </membership>
    <roleManager>
      <providers>
        <add connectionStringName="aspnetdb" applicationName="fba" name="RoleManage
Version=2.0.3600.0, Culture=neutral, PublicKeyToken=b03f5f7f11d50a3a" />
      </providers>
    </roleManager>
  </system.web>
  <connectionStrings>
    <add name="aspnetdb" connectionString="Data Source=sp2013;Initial Catalog=aspne
  </connectionStrings>
</configuration>
```

Figure 9-13: *web.config*

26. Now open web application's web.config file. Locate **<membership defaultProvider="I"><providers>** and add following snippet after tag for provider type **SPClaimsAuthMembershipProvider**.

```
<add connectionStringName="aspnetdb" applicationName="fba"
name="External" type="System.Web.Security.SqlMembershipProvider,
System.Web,&#xD;&#xA; Version=2.0.3600.0, Culture=neutral,
PublicKeyToken=b03f5f7f11d50a3a" />
```

Listing 9-2: *Connection String*

Following is the screenshot of inserted snippet:

Figure 9-14: *connectionstring in web.config*

27. Now locate **<roleManager defaultProvider=**. This will be immediately under the snippet you added above. Add following snippet in the **<providers>** tag.

```
<add connectionStringName="aspnetdb" applicationName="fba"
name="RoleManager" type="System.Web.Security.SqlRoleProvider,
System.Web,&#xD;&#xA; Version=2.0.3600.0, Culture=neutral,
PublicKeyToken=b03f5f7f11d50a3a" />
```

Listing 9-3: *Add RoleManager*

Following is the screenshot of inserted snippet:

Figure 9-15: *role manager setting in web.config*

28. Locate **</microsoft.identityModel>** tag and insert following snippet between **</microsoft.identityModel>** and **</configuration>** tags.

```
<connectionStrings>
   <add name="aspnetdb" connectionString="Data
Source=sp2013;Initial Catalog=aspnetdb;Integrated Security=SSPI" />
  </connectionStrings>
```

Listing 9-4: *Connection Strings*

Following is the screenshot of inserted snippet:

```
      </federatedAuthentication>
    </service>
  </microsoft.identityModel>
  <connectionStrings>
    <add name="aspnetdb" connectionString="Data Source=sp2013;Initial Catalog=aspnetdb;Integrated Security=
  </connectionStrings>
</configuration>
```

Figure 9-16: *Identity Model*

29. Locate **<add key="AspNetSqlMembershipProvider" value="%" />** and insert following snippet after this tag:

```
<add key="Ext" value="%" />
```

Listing 9-5: *Add key*

Inserted snippet looks like the following:

```
</SafeControls>
<PeoplePickerWildcards>
  <clear />
  <add key="AspNetSqlMembershipProvider" value="%" />
  <add key="Ext" value="%" />
</PeoplePickerWildcards>
<WorkflowServices>
  <WorkflowService Assembly="Microsoft.SharePoint, Versio
```

Figure 9-17: *sql membership*

Save both files and reset IIS. This was manual editing of web.config files. Now let me show you how you can perform same tasks using a tool.

30. Download FBA Configuration Manager for SharePoint 2013 from the following location:

http://blogs.technet.com/b/speschka/archive/2012/07/28/fba-configuration-manager-for-sharepoint-2013.aspx

31. Unzip the package to a location that is easily accessible, for example, C:\software. Package has three files:

- FBAConfigFeature.wsp
- FBAConfigMgr.exe
- HowToUseIt.txt

HowToUseIt.txt contains the installation instructions. Open **All Programs > Microsoft SharePoint 2013 Products > SharePoint 2013 Management Shell** and run following PowerShell commands:

```
add-spsolution -LiteralPath "C:\software\FBAConfigFeature.wsp"

install-spsolution -Identity fbaconfigfeature.wsp -GACDeployment
```

Listing 9-6: *Add solution*

32. Run **FBAConfigMgr.exe**.

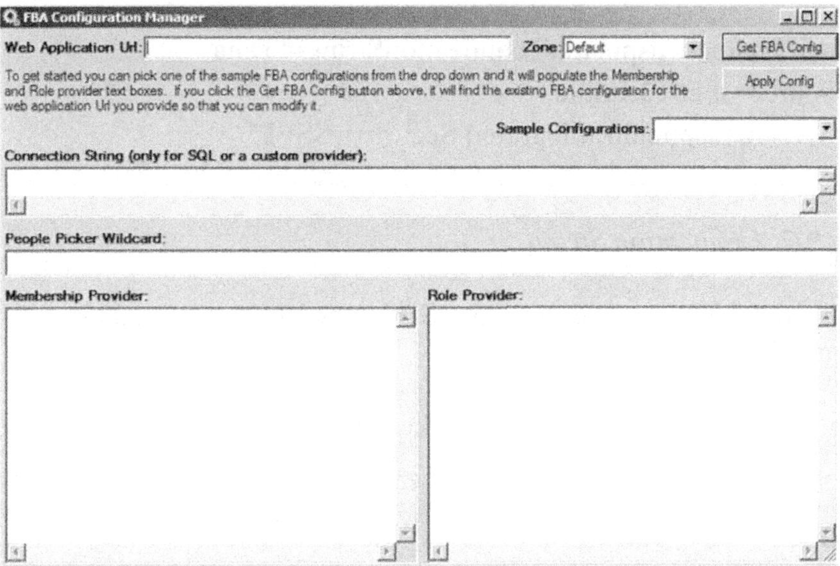

Figure 9-18: *FBA Configuration Manager*

33. In **Web Application Url**, enter the web application URL. Keep **Zone** value set to **Default** unless you chose another zone (in AAM configuration) for your web application.

34. From **Sample Configurations** dropdown, choose **SQL Connection String**. This will add a sample connection string in the **Connection String** box.

35. Again, choose **People Picker Wildcard** from the **Sample Configurations** dropdown. This will add a sample entry in the **People Picker Wildcard** box.

36. Next, choose **SQL Member** from the **Sample Configurations** dropdown. This will add a sample entry in the **Membership Provider** box.

37. Choose **SQL Role** from the **Sample Configurations** dropdown. This will add an entry in the **Role Provider** box.

38. Now, it's time to edit the sample entries.

    a. Replace value in **Connection String** box with the following:

```
<add name="aspnetdb" connectionString="Data
Source=sp2013;Initial
Catalog=aspnetdb;Integrated Security=SSPI" />
```

Listing 9-7: *Connection String*

aspnedb is the database name. **connectionString** points to the database server that hosts the aspnetdb database.

b. Replace value in **People Picker Wildcard** with the following:

```
<add key="Ext" value="%" />
```

Listing 9-8: *Add Key*

Ext is the name of the SQLMembershipProvider.

c. Replace value in **Membership Provider** with the following:

```
<add connectionStringName="aspnetdb" applicationName="fba"
name="Ext" type="System.Web.Security.SqlMembershipProvider,
System.Web,&#xD;&#xA; Version=2.0.3600.0, Culture=neutral,
PublicKeyToken=b03f5f7f11d50a3a" />
```

Listing 9-9: *Membership Provider*

d. Replace value in **Role Provider** with the following:

```
<add connectionStringName="aspnetdb" applicationName="fba"
name="ExtRole" type="System.Web.Security.SqlRoleProvider,
System.Web,&#xD;&#xA; Version=2.0.3600.0, Culture=neutral,
PublicKeyToken=b03f5f7f11d50a3a" />
```

Listing 9-10: *Role Provider*

39. Click **Apply Config** button to apply these changes to the web.config file. Please take back up of your web.config file before applying these changes. These changes will be applied to the appropriate web.config files on each server in the farm. If you ever need to make any changes to this configuration, you can use the **Get FBA Config** button to retrieve the settings from the web.config file. Make changes and click the **Apply Config** button to apply the changes back to the web.config files.

40. Now, open the SharePoint site in the browser to test everything we have configured.

You will see **Sign In** page with a dropdown box that lets you choose an authentication method.

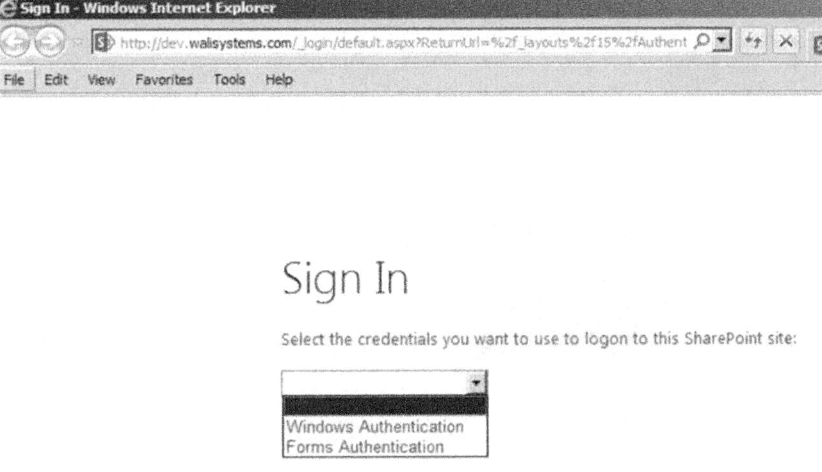

Figure 9-19: *Sign In screen*

41. Choose **Forms Authentication**. It will take you to the login screen. Enter user name and password for one of the users you created above in the external database and click **Sign In** button.

Warning: this page is not encrypted for secure communication. User names, passwords, and any other information will be sent in clear text. For more information, contact your administrator.

User name: saif

Password: ••••

Sign In

Sign me in automatically

Figure 9-20: *Login screen*

42. This is how the extranet site looks with FBA configured.

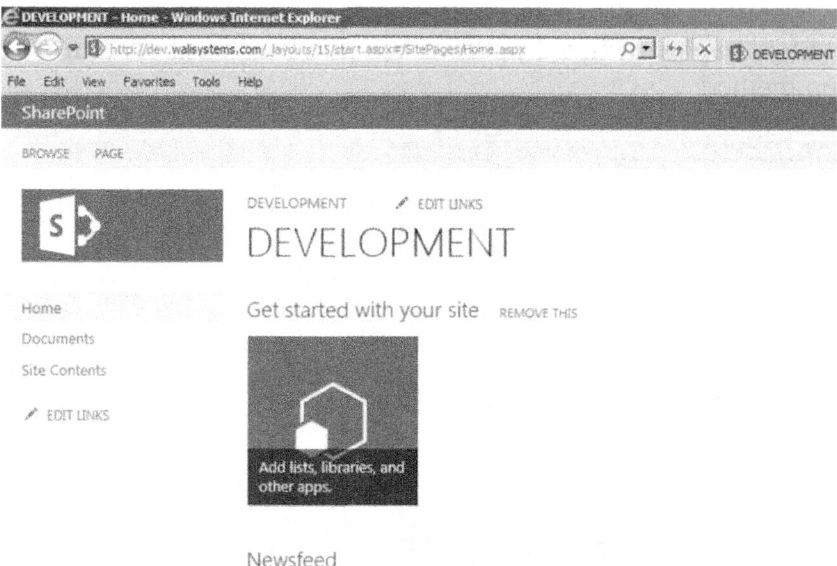

Figure 9-21: *Extranet site*

Summary

There are different options for setting up Extranets. One can use LDAP, one can use custom database, or an OpenID. In this chapter, you learned to use custom database. By default, authentication type in SharePoint 2013 is claims-based. To access custom database or data source like LDAP (Lightweight Directory Access Protocol), forms-based authentication has to be configured in web application settings. Forms-based authentication is an identity management system that is based on ASP.NET membership and role provider authentication.

Chapter 10

SharePoint and HTML 5

In this chapter:

- You will learn to create pages using HTML5 code.
- You will create a video player in HTML5.
- You will learn tips and techniques to make HTML5 video run in IE9 (and newer versions).
- You will learn to use custom HTML5 video player in SharePoint 2013.

Are you excited? Have you been hearing this buzz word (HTML5) for a long time but couldn't muster enough confidence or energy to venture into it. Well, guess what! Now is the time! If you have ever wanted to learn this technology, you are at the right place. Just imagine, HTML5 and SharePoint 2013! Together! That's seems unbelievable! Isn't it? Well, I will be honest here. It's not about SharePoint. The problem is with the browsers. If you write HTML5 code and if you can run it in the browser, you can run it inside SharePoint as well but if you cannot run it in the browser, how will you run it inside SharePoint! Different browsers treat HTML5 differently. HTML5, if you talk about syntax, is just like ordinary HTML. What make it exciting are the new tags. It just doubles the amount of work that programmers have to do to make the code compatible with all browsers. I must tell you that IE9 lags far behind the other browsers. The latest IE version available on the market at this time is IE11. Since I am a consultant, I work with all types of clients, all major enterprises and I can tell you that most big companies are still using IE8 and the reason is its stability. IE9 hasn't been a big success. It had flaws and it rarely worked with the latest technologies. If you can make something work in IE9, then that stuff will definitely work in the newer versions such as IE10 and IE11 because some of the flaws and compatibility issues have been fixed in the newer versions. It excited many people when it was announced that IE9 would support HTML5. No doubt, IE9 supports HTML5 but there are several things that you have to do (configure manually) before you can take advantage of the tags that have induced so much excitement in the developer community. Here are few things that you can do with HTML5:

1. You are not dependent on Flash anymore. You can create rich Internet applications with HTML5. Not only that, you can now provide same rich functionality with simple code (no clutter, no mess).

2. You can create rich applications for both web and mobile devices. Interesting thing is that you can now create an application that will run in mobile devices without the need to create native application. If you were afraid of programming in objective C, now you can create apps using just HTML(5). Cool! Isn't it?

3. Create animations using Canvas. Using HTML5 canvas, you can now create games and other animations with ease.

4. You can create amazing applications by using HTML5 with Javascript. You can take advantage of different Javascript libraries that are available for free. One of them is JQuery. You can create cool apps with HTML5 and JQuery.

5. Geolocation is a cool new feature that lets you find out where in the world your visitors are. Luckily, geolocation feature has good support in every browser.

6. Audio and Video: Adding audio and video to your web pages is a breeze now. Just add <audio> tag for audio and <video> tag for video and you are done. Add source tag to define the path and your media-rich page is ready.

7. Simplicity: Now you can create useful complex applications with ease. All you need is HTML and some Javascript and you are ready to rock.

With the help of Javascript programming, you can create your own controls to control the video. This means you can create your own video player, apply different styles to it, add the functionality you want your way. Controls attribute in the video tag allows you to use the default controls in the video player. If you remove this attribute, all you get is a video window (without any controls). You can add your own controls underneath the video window.

1 2 3 Go!

As I said, you can create rich Internet applications with HTML5 and since HTML5 has native support in SharePoint 2013, bringing the two technologies together to develop useful applications, will add value to SharePoint. SharePoint used to be a portal technology, now it's a full-fledged development environment. You can create custom applications of any kind for your audiences. With the advent of SharePoint

Online, the usefulness of SharePoint has increased many folds. Without the expensive infrastructure and hardware, you can run your sites in the cloud. Up until now, SharePoint was a tool of choice for developing organization-wide Intranets but with SharePoint Online, organizations can setup Extranets and Public facing Internet sites very easily. On top of that, if you want to spice-up your public facing sites, use HTML5. Great thing is that HTML5 works with other popular tools, for example, JQuery. By using JQuery/Javascript with HTML5, you can develop awe-inspiring applications and host them in SharePoint. Since this book focuses on SharePoint, we will not discuss HTML5 in detail but I will show you how you can write HTML5 code for SharePoint. You will learn to use some of the most popular HTML5 tags, for example, geolocation and video tags.

HTML5 Geolocation and SharePoint

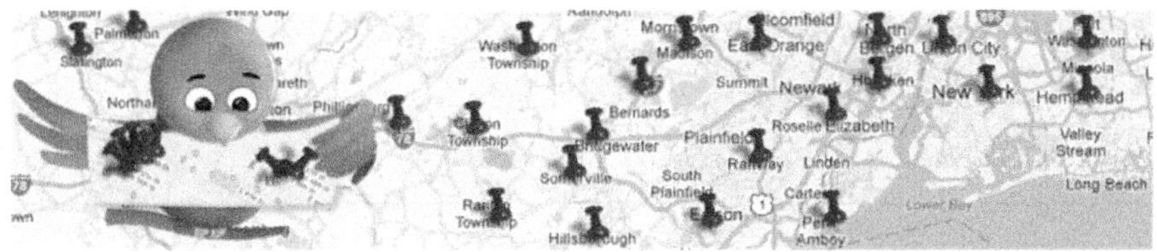

Figure 10-1: *HTML5 geolocation and mobile devices*

Let's me show you how you can take advantage of HTML5 Geolocation feature in SharePoint. Geolocation allows you to find the location of your site visitors. There can be many uses of this feature. For example, you can use it to gather statistics for your site. You can then create a graph or chart widget and show countries that visited your site or blog the most. There are endless possibilities. You can also show cities. It can really come in handy if you run a business on the net. This thing also works in mobile devices. Looking at the number of mobile applications that are sprouting every day, this feature can really play a pivotal role in gathering geographic data of your visitors. It will directly impact your sales and marketing campaigns. Plus point is that this feature has good support in every browser. Well if you are thinking about IE 7 and IE 8, forget it. You need at least IE 9 for this thing to work but that's for computers. For devices that use non-IE browsers, this isn't a problem. Almost all browsers support this feature. If you are afraid that this thing will not work in SharePoint, let me tell you that it will work. All you need is IE9 browser. If you don't have IE 9, start using Chrome or Firefox. They work well with SharePoint 2013 and also support other standards that are popular today. This includes HTML5 technology.

Let's start with coding first. Once I have explained the code, I will show you how to add that code to a SharePoint page.

Geolocation

Using Geolocation is very easy. Check out the following piece of code:

```
navigator.geolocation.getCurrentPosition(
function(position)
{
alert("Your location is: " + position.coords.latitude + ", " +
position.coords.longitude);
}<BR>);
```

Listing 10-1: *Use of Geolocation*

You get latitude and longitude coords to find the position of the user. OK, now you know how to find the location. Next goal is to show this location on the map. We will use Google Maps for this purpose. Again, it's not difficult to use Google Maps. If you haven't used Google Maps before, you may want to check out the documentation online at http://code.google.com/apis/maps/documentation/javascript. You have to include a Google Map link in your code before you can actually use this service. Add following line between the <head> ... </head> tags.

```
<head>
<title>HTML5Geolocation</title>
<scriptsrc="http://maps.google.com/maps/api/js?sensor=true"></script>
...< BR></head>
```

Listing 10-2: *Calling Google maps API*

Add a div tag to the page. This will be used to draw the map.

```
<body>
<div id="mapCanvas"></div>
```

Listing 10-3: *Add a div tag to draw the canvas*

Use following style to design the map layout:

```
#mapCanvas
{
width: 650px;
height: 450px;
border: solid 1pxblue;
}
```

Listing 10-4: *Map Canvas Style*

Now, you are ready to see everything in action. Call following function when the window loads:

```
window.onload = function() {
results = document.getElementById("results");
var myOptions = {
zoom: 13,
mapTypeId: google.maps.MapTypeId.ROADMAP
};
```

Listing 10-5: *Calling main function*

This code writes results in the "results" div. Variable myOptions contains options like zoom level and the map type. Map type that we are going to use is ROADMAP. After setting the options, create the map:

```
map = new google.maps.Map(document.getElementById("mapCanvas"),
myOptions);
```

Listing 10-6: *Create the map*

Following code will set user's location on the map:

```
function geolocationSuccess(position) {

var location = new
google.maps.LatLng(position.coords.latitude,position.coords.longitude);
map.setCenter(location);

var infowindow = new google.maps.InfoWindow();

infowindow.setContent("This is your location on the map!");
infowindow.setPosition(location);
infowindow.open(map);

results.innerHTML = "You are on the map.";
}
```

Listing 10-7: *Set user's location on the map*

Variable location contains the exact location. Then you show a box on the map with a message. You use InfoWindow() function to show the box. This box will appear at user's location. The following function will be called in case of failure:

```
function geolocationFailure(positionError) {
if (positionError.code == 1) {
results.innerHTML =
"You have decided not to share your location.";
}
else if (positionError.code == 2) {
results.innerHTML =
"The network is down or the positioning service can't be reached.";
```

```
}
else if (positionError.code == 3) {
results.innerHTML =
"The attempt timed out before it could get the location data.";
}
else {
results.innerHTML =
"Unknown error. Please try later.";
}
}
```

Listing 10-8: *Error handling*

In geolocationFailure() function you check different conditions to know the cause of failure and then show message accordingly. That's it.

You can download complete code from the book's website.

Now you have the code. All you have to do is use it inside a SharePoint page.

1. Open SharePoint site in SharePoint Designer. Click on **Site Pages** from the **Navigation**.

2. Click **Page** button (click the arrow to open the submenu) in the ribbon and select **ASPX**.

3. Replace **Untitled_1** with **geolocation** to rename the file. The file name will be geolocation.aspx.

4. Right-click geolocation.aspx and select **Check out**. Again right-click and select **Edit File in Advanced Mode**.

5. Locate **<title>** tag and change the title from **Untitled_1** to **GeoLocation Map**.

6. Add following code after the **</title>** tag.

```
<script
src="http://maps.google.com/maps/api/js?sensor=true"></script>
<Style type="text/css">
```

```
#mapSurface
{
    width: 600px;
    height: 400px;
    border: solid 1px black;
}
</style>
```

Listing 10-9: *Call Google maps API and add style sheet*

7. Remove **<form> … </form>** tags from the **<body>** and add following code between **<body>** and **</body>** tags.

```
<p id=results>where do you live?</p>
<div id="mapSurface"></div>
<script type="text/javascript">
var results;
var map;
window.onload = function()
{
     results = document.getElementById("results");
    var myOptions = {
        zoom: 13,
        mapTypeId: google.maps.MapTypeId.ROADMAP
    };

    //create the map, with these options.

    map = new google.maps.Map
(document.getElementById("mapSurface"), myOptions);

    //Try to find the visitor's position
    if (navigator.geolocation)

    {

    navigator.geolocation.getCurrentPosition(geolocationSuccess,
geolocationFailure);
    results.innerHTML = "The search has begun.";
```

```
    }
    else
    {
            results.innerHTML = "This browser does not support 
geolocation.";
            goToDefaultLocation();
    }
};
function geolocationSuccess(position)
{

    location = new google.maps.LatLng(position.coords.latitude, 
position.coords.longitude);

    map.setCenter(location);
    var infowindow = new google.maps.InfoWindow();
    infowindow.setContent("You are here");
    infowindow.setPosition(location);
    infowindow.open(map);
    results.innerHTML = "Now you are on the map";
}
function geolocationFailure(positionError)
{
    goToDefaultLocation();
}
function goToDefaultLocation()
{
    var newYork = new google.maps.LatLng(40.69847, -73.95144);
    map.setCenter(newYork);
}
</script>
```

Listing 10-10: *Complete code*

8. **Save** the page.

9. Right-click the **geolocation.aspx** under **Site Pages** (in left navigation) and select **Check In**. Enter comments and click **OK**.

10. Right-click the **geolocation.aspx** and select **Preview in Browser**.

11. You will be prompted to allow the browser to access your computer's location. Allow the access, map will appear immediately showing your location.

12. You can also show this map on any of your existing SharePoint pages by using a web part. You can use **Page Viewer** web part so access the page you created above. Add **Page Viewer** web part to the page where you want the map to appear, edit web part properties. Make sure **Web Page** radio option is selected, add **/SitePages/geolocation.aspx** to the Link box and click **OK**. Re-load the page and you will see the map on the page.

    Another way of doing the same thing is to use **Content Editor** web part. This is really helpful if you have not created a separate ASPX page and want to put code directly inside the web part. Add **Content Editor** web part on the page, click inside the web part and then select **Edit Source** button from the ribbon and then copy the complete code provided in the listing above in the web part source area. Save the changes. Re-load the page and bingo! Page shows the map inside the web part. You can position the web part anywhere you like on the page.

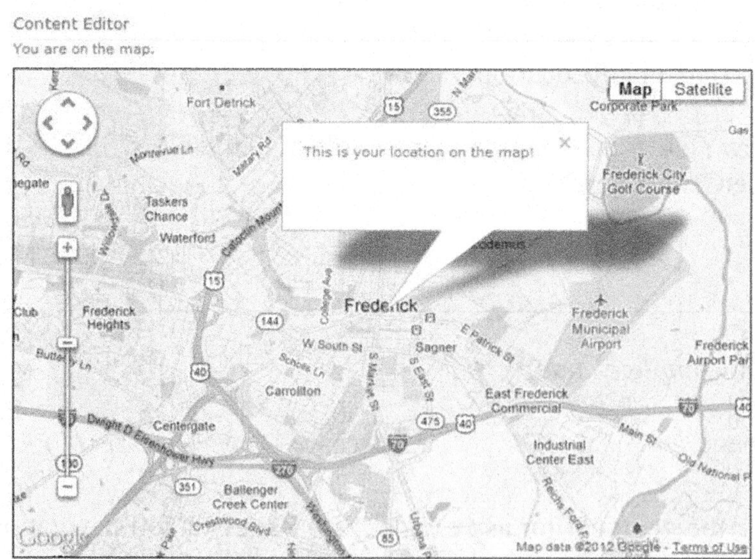

Figure 10-2: *Geolocation map inside a content editor web part*

If your site is a public website, you can preview the page on a mobile device.

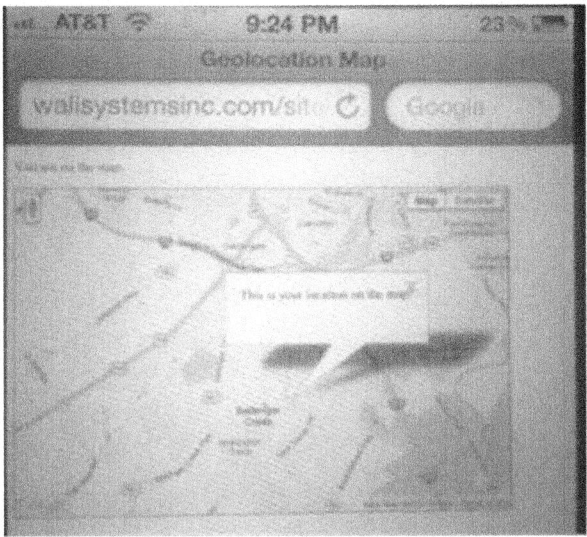

Figure 10-3: *Geolocation map on a mobile device*

Twitter, Geolocation and SharePoint

In this section, we will extend the idea further and include Twitter. In the last section, you saw how one can find location of site visitors using HTML5 Geolocation feature. Now we will use that feature to find Twitter users in your area. We will add two text boxes in the application. One to enter the search term and the other to enter the radius of the search. Application will first find your location on the map and then search in the proximity of your location. You can set radius to any value (in miles) you like. Application will find Twitter users in that radius of your location and then show their logo, name, tweet and tweet time. It works in browsers as well as mobile devices.

1. Open SharePoint site in SharePoint Designer. Click on **Site Pages** from the **Navigation**.

2. Click **Page** button (click the arrow to open the submenu) in the ribbon and select **ASPX**.

3. Replace **Untitled_1** with **findtwitterusers** to rename the file. The file name will be geolocation.aspx.

4. Right-click geolocation.aspx and select **Check out**. Again right-click and select **Edit File in Advanced Mode**.

5. Locate **<title>** tag and change the title from **Untitled_1** to **Twitter Search Using HTML5 GeoLocation**.

6. Add following code before the **</head>** tag.

```
<style type="text/css">
    td {
        padding-top: 15px;
        padding-right: 15px;
        padding-bottom: 20px;
        padding-left: 15px;
        border: 1px solid black;
    }
    table {

        text-align: center;
        border-collapse: collapse;
    }

    a:link {
        text-decoration: none;
        color: #0000ff;
    }

    .msg {
        text-align: left;
        padding-left: 8px;

    }

    tr.odd {
        background-color: #effbff;

    }

    tr.even
    {
        background-color: #fff8b2;
```

```
    }

.src {
        text-align: center;
        font-style: normal;
        padding: 10px;

    }

</style>

<script type="text/javascript">

    function $(id){
            return document.getElementById(id);

    }

    function twitterSearch(){
            var gps = navigator.geolocation;
            if (gps){
                    gps.getCurrentPosition(searchTweets,showError);

            } else {
                    searchTweets();

            }
    }

    function showError(error)
    {
            //Display error message
            alert("Error occurred! Error Code: " + error.code + " Error
Message: " + error.message);

    }

    function searchTweets(position){

            var query =
"http://search.twitter.com/search.json?callback=searchResults&q=";
```

```
            query += document.getElementById("searchBox").value;

            if (position){
                    var lat = position.coords.latitude;
                    var lng = position.coords.longitude;
                    query += "&geocode=" + escape(lat + "," + lng + "," +
document.getElementById("radiusBox").value + "mi");

            }

            var script = document.createElement("script");
            script.src = query;

    document.getElementsByTagName("head")[0].appendChild(script);

    }

    function searchResults(response){
            var tweets = response.results;
            tweets.forEach(function(tweet){
            tweet.linkUrl = "http://twitter.com/#!/" + tweet.from_user +
"/status/" + tweet.id;

            });

            createTable(tweets);

    }

    function createTable(tweets){
            var counter = 1;
            var rows = tweets.map(function(tweet){
            counter++;

            return createTableBody(tweet.from_user,
tweet.profile_image_url, tweet.text,        tweet.linkUrl, tweet.created_at,
tweet.geo, counter % 2 == 0);

            });

            document.getElementById("searchResults").innerHTML =
```

```
createTableSkeleton();

            rows.forEach(function(row){

    document.getElementById("resultsTable").appendChild(row);

            });
    }

    function createTableSkeleton()
    {

            var tbl = "<table id='resultsTable' border='1'
bgcolor='lightskyblue'>";

            tbl += "<TR><TH><h3>Logo</h3></TH>";
            tbl += "<TH><h3>User</h3></TH>";
            tbl += "<TH><h3>Tweet</h3></TH>";
            tbl += "<TH><h3>Tweeted
At</h3></TH></TR><BR><BR>";
            tbl += "</table>";

            return tbl;

    }

    function createTableBody(userID, logoUrl, tweetText, linkUrl,
created_At, geo, IsAlternateRow)
    {

      var resultsRow = document.createElement("tr");
      if (IsAlternateRow){
        resultsRow.setAttribute("class", "odd");

      }

      Else
      {

            resultsRow.setAttribute("class", "even");
```

```
        }

        var srcCell = document.createElement("td");
        srcCell.setAttribute("class","src");
        resultsRow.appendChild(srcCell);
        var logo = document.createElement("img");
        logo.src = logoUrl;
        logo.setAttribute("alt", userID);
        logo.setAttribute("height", 48);
        logo.setAttribute("width", 48);

        srcCell.appendChild(logo);

        var user = document.createElement("td");
        user.setAttribute("class","src");

        resultsRow.appendChild(user);

        user.appendChild(document.createTextNode(userID));

        var messageCell = document.createElement("td");
        messageCell.setAttribute("class","msg");

        var link = document.createElement("a");
        link.setAttribute("href", linkUrl);
        link.setAttribute("target", "_blank");
        link.appendChild(document.createTextNode(tweetText));

        messageCell.appendChild(link);

        resultsRow.appendChild(messageCell);
        var created_at = document.createElement("td");

        created_at.setAttribute("class","src");

        resultsRow.appendChild(created_at);

    created_at.appendChild(document.createTextNode(created_At));

    return resultsRow;
```

```
        }

    </script>
```

Listing 10-11: *Complete code*

7. Remove **<form></form>** tags from the body and copy following code between the **<body>** and **</body>** tags.

```
<div id="main">
                <label for="searchBox">Enter Search
Term:</label>      
                <input type="text" id="searchBox"/><br>
                <label for="radiusBox">Enter Radius (in miles):</label>
                <input type="text" id="radiusBox"/>
                <input type="button" value="Search Twitter"
onclick="twitterSearch()"/>
        </div>
        <div id="searchResults">
        </div>
```

Listing 10-12: *Add controls to the page*

7. Save the page and preview in the browser.

8. Enter a search term and radius and click **Search Twitter** button.You will get a message that “server wants to access the physical location”. Click **Allow**. The page will show you the results.

Code Explanation

```
function twitterSearch(){

var gps = navigator.geolocation;
if (gps){
    gps.getCurrentPosition(searchTweets,showError);
```

```
    }
    else
    {
    searchTweets();
    }
}
```

Listing 10-13: *twitterSearch() function*

This function gets the current position of the user. searchTweets() is the function that is called if current location is found successfully. searchTweets() is next, shown below.

```
function searchTweets(position){

var query =
"http://search.twitter.com/search.json?callback=searchResults&q=";
query += document.getElementById("searchBox").value;
if (position){
var lat = position.coords.latitude;
var lng = position.coords.longitude;
query += "&geocode=" + escape(lat + "," + lng + "," +
document.getElementById("radiusBox").value + "mi");
}

var script = document.createElement("script");
script.src = query;
document.getElementsByTagName("head")[0].appendChild(script);
}
```

Listing 10-14: *searchTweets() function*

searchResults() is the callback function. We pass search keyword in the query parameter called "q" and then we pass geocode that comprises of latitude, longitude and radius. Then we add this search query dynamically to the page body. searchResults() function is given below:

```
function searchResults(response){
var tweets = response.results;
tweets.forEach(function(tweet){
tweet.linkUrl = "http://twitter.com/#!/" + tweet.from_user + "/status/" +
tweet.id;
});
createTable(tweets);
}
```

Listing 10-15: *searchResults() function*

In searchResults() we read each tweet and create results table. We pass all tweets to the createTable() function. createTable() function is given below:

```
function createTable(tweets){
var counter = 1;
var rows = tweets.map(function(tweet){
counter++;
return createTableBody(tweet.from_user, tweet.profile_image_url,
tweet.text,
tweet.linkUrl, tweet.created_at, tweet.geo, counter % 2 == 0);
});

document.getElementById("searchResults").innerHTML =
createTableSkeleton();
rows.forEach(function(row){
document.getElementById("resultsTable").appendChild(row);
});
}
```

Listing 10-16: *createTable() function*

createTableSkeleton() function creates the table skeleton and header row.

```
function createTableSkeleton()
{
var tbl = "<table id='resultsTable' border='1' bgcolor='lightskyblue'>";
```

```
tbl += "<TR><TH><h3>Logo</h3></TH>";
tbl += "<TH><h3>User</h3></TH>";
tbl += "<TH><h3>Tweet</h3></TH>";
tbl += "<TH><h3>Tweeted At</h3></TH></TR><BR><BR>";
tbl += "</table>";

return tbl;
}
```

Listing 10-17: *createTableSkeleton() function*

createTableBody() creates the main body that shows the results.

```
function createTableBody(userID, logoUrl, tweetText, linkUrl, created_At,
geo, IsAlternateRow)
{

var resultsRow = document.createElement("tr");
if (IsAlternateRow){
resultsRow.setAttribute("class", "odd");
}
else
{
resultsRow.setAttribute("class", "even");
}
var srcCell = document.createElement("td");
srcCell.setAttribute("class","src");
resultsRow.appendChild(srcCell);

var logo = document.createElement("img");
logo.src = logoUrl;
logo.setAttribute("alt", userID);
logo.setAttribute("height", 48);
logo.setAttribute("width", 48);
srcCell.appendChild(logo);

var user = document.createElement("td");
user.setAttribute("class","src");
resultsRow.appendChild(user);
```

```
user.appendChild(document.createTextNode(userID));


var messageCell = document.createElement("td");
messageCell.setAttribute("class","msg");
var link = document.createElement("a");
link.setAttribute("href", linkUrl);
link.setAttribute("target", "_blank");
link.appendChild(document.createTextNode(tweetText));
messageCell.appendChild(link);
resultsRow.appendChild(messageCell);


var created_at = document.createElement("td");
created_at.setAttribute("class","src");
resultsRow.appendChild(created_at);
created_at.appendChild(document.createTextNode(created_At));


return resultsRow;
}
```

Listing 10-18: *createTableBody() function*

Then there is this style sheet. It adds some styling to the controls and the search results, like alternating colored rows, photo thumbnails, etc.

```
<style type="text/css">
td {
padding-top: 15px;
padding-right: 15px;
padding-bottom: 20px;
padding-left: 15px;
border: 1px solid black;
}
table {
text-align: center;
border-collapse: collapse;
}
```

```
a:link {
text-decoration: none;
color: #0000ff;
}
.msg {
text-align: left;
padding-left: 8px;
}
tr.odd {
background-color: #effbff;
}
tr.even
{
background-color: #fff8b2;
}
.src {
text-align: center;
font-style: normal;
padding: 10px;
}
</style>
```

Listing 10-19: *Style sheet*

Finally we have the html controls. Results are shown in "searchResults" DIV.

```
<div id="main">
<label for="searchBox">Enter Search
Term:</label>      
<input type="text" id="searchBox"/><br>
<label for="radiusBox">Enter Radius (in miles):</label>
<input type="text" id="radiusBox"/>
<input type="button" value="Search Twitter" onclick="twitterSearch()"/>
</div>
<div id="searchResults">
</div>
```

Listing 10-20: *Controls*

This is how it will look in SharePoint:

Enter Search Term: sharepoint
Enter Radius (in miles): 100 [Search Twitter]

Logo	User	Tweet
	sptrainingcente	SharePoint Training Center Daily is out! http://t.co/Ea2tc4Ya ▸ Top stories today via @dbargenis @l
	suresh_kothari	RT"@yourjewishnews: Israel to force hospitalization of people suffering from anorexia http://t.co/dC
	drlbend	RT @yourjewishnews: California raises taxes to fund free condoms for teens http://t.co/ULLyJXsr
	HerwinWevers	@svenblom ja. Had ik gezien. En ben het met je eens. Bouwen van #zs op sharepoint kost flink wat t

Figure 10-4: *Twitter results in SharePoint*

If your site is a public website, you can view the page on a mobile device. You will be prompted to allow the device to access your current location. Click **OK**. Your device will show you the Twitter users in your area tweeting about the keyword you entered in the search box.

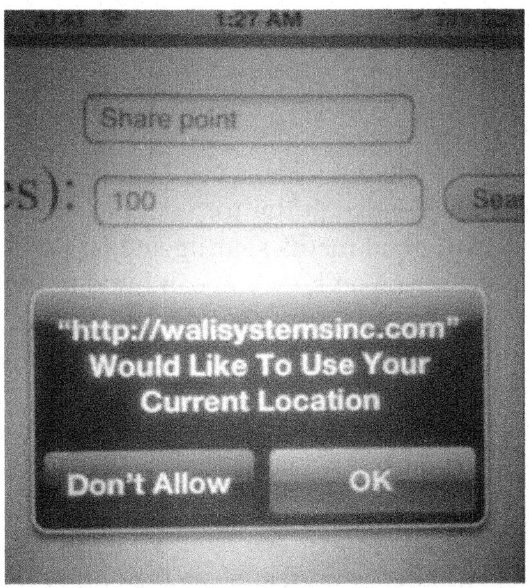

Figure 10-5: *Searching Twitter users on a mobile device*

You can also show this application on any of your existing SharePoint pages by using a web part. You can use **Page Viewer** web part so access the page you created above. Add **Page Viewer** web part to the page where you want the controls to appear, edit web part properties. Make sure **Web Page** radio option is selected, add **/SitePages/findtwitterusers.aspx** to the Link box and click **OK**. Re-load the page and you will see the map on the page.

Another way of doing the same thing is to use **Content Editor** web part. This is really helpful if you have not created a separate ASPX page and want to put code directly inside the web part. Add **Content Editor** web part on the page, click inside the web part and then select **Edit Source** button from the ribbon and then copy the complete code provided in the listing above in the web part source area. Save the changes. Re-load the page and bingo! Page shows the map inside the web part. You can position the web part anywhere you like on the page.

You can download complete code from the book's website.

HTML5 Video and SharePoint

SharePoint 2013 has a built-in HTML5 media player and it can play all video files that are compatible with the HTML5 <video> implementation for the current browser. You can now play videos on mobile devices that do not support Silverlight. If the format is not playable by the HTML5 player, then we use other tools.

SharePoint 2013 offers a video player page for each video. Each page has metadata for the video and all the properties that user fills in. Videos can be played from the video player page. This allows you to build a video portal for your organization. This can also come in handy if you build a social media sharing site on SharePoint 2013. The branding chapter in this book shows you how you can build a public facing website in SharePoint 2013 and in that chapter we build a site called "Social Playground". It has a slider widget on the home page. You can replace that slider widget with the Media Web part in SharePoint or you can keep the slider and also include Media Web Part to show videos on the site. Sky is the limit. You can implement as many ideas as you like.

Another great feature is that users can download the video from your site, or they can embed the video in a different site. Another good feature is the ability to add related items. You can drag and drop the relevant files on the same page that hosts the video.

Using Media Web Part

You can upload videos to any library but the preferred location is the library that has the video content type enabled. For example, the Asset library, which is customized for storing digital assets.

Uploading a video is very simple. Use drag and drop to upload files. This is very useful if you want to upload large number of files.

You can also upload video from your computer and host it inside SharePoint using a Media Web Part.

1. To upload a video, go to the page where you want to show the video and click **Edit** to edit the page.

2. From the **INSERT** tab, go to **Video and Audio** button in the ribbon and click the arrow to see the options. You see following four options:

- From Computer
- Embed
- From SharePoint
- From Address

3. Select **From Computer**, click **Browse** button to locate the video file on your computer. For testing, the book provides a video file (HTML5_Video_Player_Demo_File.mp4) that you can download from the book's website. Choose **Destination Library** from the drop down. You can either choose **Images** or **Site Assets** library. Click **OK**. That's it. You can now play the video. You can enlarge the screen by clicking the button in the top right corner of the video.

Figure 10-6: *Media Web part in SharePoint 2013*

To explore more video capabilities, add an **Asset Library** application to your site.

1. Go to **Site Actions > Site Contents** and select **Asset Library**.

2. In the **Name** field, enter **My Digital Assets**. Click **Create**.

3. Click **new item** link to add video to the library. Click **Browse** button to locate the video file. Click **OK**.

4. On the page that opens, you can fill in the value for **Description** field. You can change the thumbnail by clicking the **Change thumbnail** link under the video. Expand **This video's renditions** section. A user can have multiple renditions of the same video. For example, the same video with different playback quality or subtitles is considered a rendition. Click **Manage Renditions** link. A new window will pop up. In this window, you can add more renditions for your video. These are the video renditions available to users when playing back the video. Multiple video renditions are useful when you want a video to stream quickly for users on slower connections, but allow users to view a higher quality version on faster connections. You can also set default video rendition. This is the video rendition that is loaded by default. You can choose your video file in this drop down or you can leave the default option selected which is **Use the video with the lowest bitrate**, in this case, video player will choose the suitable video itself.

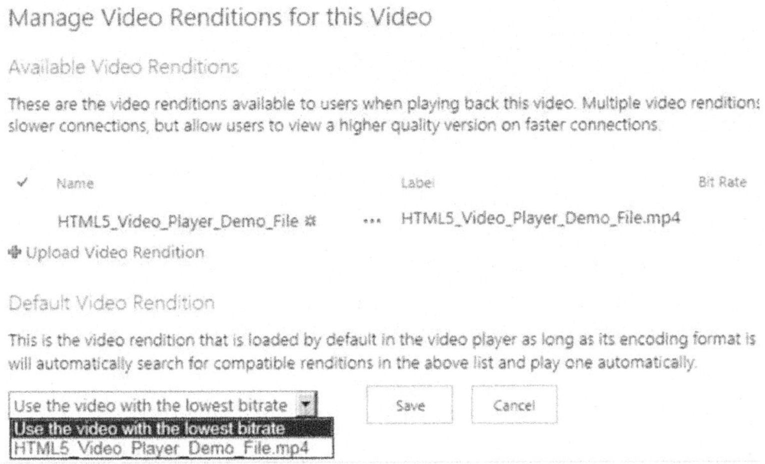

Manage Video Renditions for this Video

Available Video Renditions

These are the video renditions available to users when playing back this video. Multiple video rendition: slower connections, but allow users to view a higher quality version on faster connections.

✓	Name		Label	Bit Rate
	HTML5_Video_Player_Demo_File	...	HTML5_Video_Player_Demo_File.mp4	

Upload Video Rendition

Default Video Rendition

This is the video rendition that is loaded by default in the video player as long as its encoding format is will automatically search for compatible renditions in the above list and play one automatically.

Use the video with the lowest bitrate
Use the video with the lowest bitrate
HTML5_Video_Player_Demo_File.mp4
Save
Cancel

Figure 10-7: *Manage video renditions*

5. Click **Save** to save the settings. You will be returned to the previous window. Click **Save**.

6. Click the video thumbnail to play the video. On the video page, on the right side, you see **Related** section that shows related files that are relevant to the users watching this video. **DOWNLOAD** lets the users download this video.

SharePoint provides two types of thumbnailing: server-generated and user-generated. Default thumbnail is auto-generated by SharePoint. It is the server-generated thumbnail. For user-selected thumbnailing, you can use SharePoint's client-side thumbnail generator. The thumbnail generator can be accessed from the video edit form. When you choose the change thumbnail link, you get an option to capture thumbnail from video, use a picture from the computer, or use a picture from a web address.

Some other cool features that are available on the video page are tagging, ratings, view count, and faster video search. You can explore these in the **Asset Library** application you added above.

You have explored the HTML5 video player that SharePoint provides out-of-the-box. In the next section, we will develop a custom video player using HTML5. The objective is to show you how you can develop HTML5 applications for SharePoint 2013. I chose to develop video player because it is a fancy application and you get to learn several new things. You can also develop fancy forms in HTML5, drag and drop applications, image rendering applications, and so on.

Prepare Machine for HTML5 Video

Before you start using the <video> tag, you need to make sure that your system and browser supports this tag. Please note, that you may already be using an up-to-date system and applying updates may not be necessary but in case you have tried and failed to run <video> tag on your system, you need to follow the steps below to make it work. I will show you different tips and techniques on how to make system compatible to work with the <video> tag. After that you will create a video player and use it in SharePoint.

Following browsers support HTML5 <video> tag:

H.264: IE (9, 10, 11), Chrome, Safari

Ogg Theora: FireFox 3.5, Chrome 5, Opera 10.5

WebM: FireFox 4, Chrome 6, Opera 10.6

Minimum version numbers that are needed for the <video> tag are mentioned above.You will notice that IE works with H.264 format only. You will have to convert your videos to this format in order to play them inside IE browser. It's so easy to play videos in other browsers but with IE it's a pain. You have to make it work forcefully.

HTML5 audio/video works with following file types:

- MP3 (.mp3)
- 2. Ogg Vorbis (.ogg)
- 3. WAV (.wav)
- 4. H.264 (.mp4)
- 5. Ogg Theora (.ogv)
- 6. WebM (.webm)

Correct MIME type should be set in IIS before you can use a particular file type in browser. Before a web server sends a resource to a browser, it sends the MIME type. For example, to run .mp4 video file, **video/mp4** MIME type should be added to your site in IIS first. As you have seen above that IE works with H.264 (.mp4) format only so it's important that you add this MIME type in IIS first. Follow the steps below to add this MIME type:

Add video/mp4 MIME type in IIS

1. Go to **All Programs > Administrative Tools** and select **Internet Information Services (IIS) Manager**.

2. You can add MIME type directly to the site that you may have created for your HTML5 project or you can add it to the server so that all sites can use it.

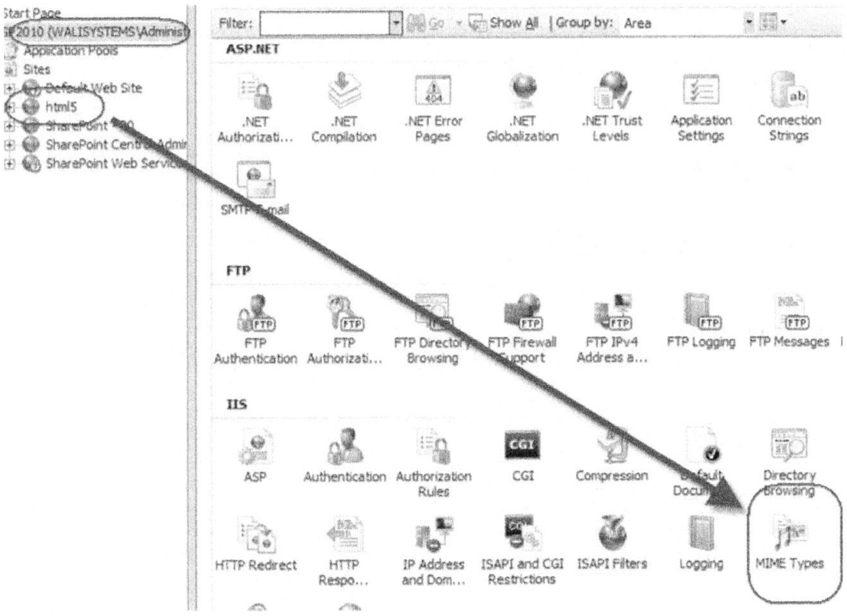

Figure 10-8 : *Add a MIME type*

I created a site for my HTML5 projects. Select the site where you want to add the MIME type and then on the right side, in **IIS** section, double-click **MIME Types** icon to open MIME types.

3. On right side, in **Actions** pane, click **Add...**link.

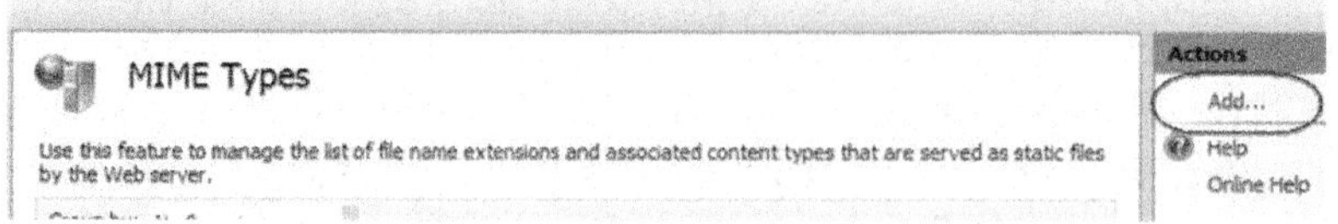

Figure 10-9: *Click Add to add new MIME type*

4. Enter **.mp4** in **File name extension** and **video/mp4** in **MIME type** and click **OK**.

Add MIME Type

File name extension:

.mp4

MIME type:

video/mp4

OK Cancel

Figure 10-10: Add file extension

A new MIME type has been added.

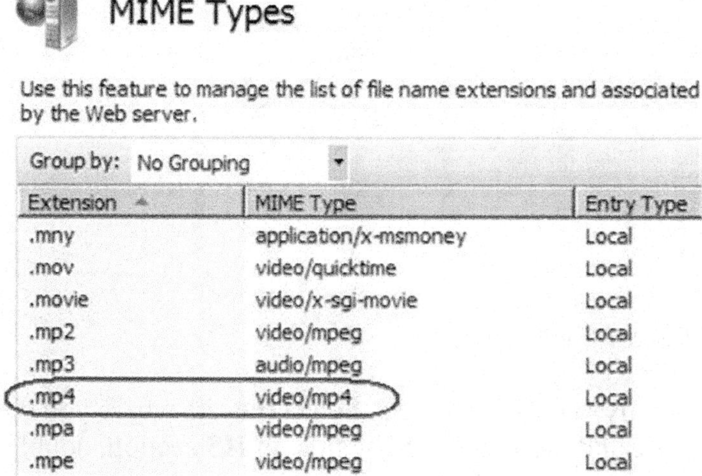

Extension	MIME Type	Entry Type
.mny	application/x-msmoney	Local
.mov	video/quicktime	Local
.movie	video/x-sgi-movie	Local
.mp2	video/mpeg	Local
.mp3	audio/mpeg	Local
.mp4	video/mp4	Local
.mpa	video/mpeg	Local
.mpe	video/mpeg	Local

Figure 10-11: *Newly added MIME type*

Install Internet Explorer 9

5. If you are still running IE8 or older browser, then download IE 9 for Windows 7 from the following link:

   http://www.microsoft.com/download/en/details.aspx?id=13950

   Download IE 9 for Vista and Windows Server 2008 from the following link:

   http://www.microsoft.com/download/en/details.aspx?id=16792

   You can also install IE 10 if you want but do not install IE 11 as yet as it is not compatible with SharePoint 2013. Microsoft may release a patch soon to fix the issue but until then stick to IE 9 or 10. Web Part properties do not show up if you open site in IE 11.

6. After you have installed IE 9, you have to make sure that it supports the HTML5 video format. So, it's time to play with the HTML5. Let's create an HTML5 page and add a video tag to it. Download sample MP4 video file from the book's website to use in your code. It's in a zip format and size is 2.41 MB. You can also play the file directly from the web without downloading it to your machine but that will be a slower experience. For good quality, it's better to download the file. If you want to run it directly from the net, here is the URL:

http://walisystemsinc.com/sharepoint/art/html5/HTML5_Video_Player_Demo_File.mp4

7. Open notepad and add following code:

```
<!DOCTYPE html>
<html>
<body>
<video id="video1" controls>
<source src="HTML5_Video_Player_Demo_File.mp4" type='video/mp4;
codecs="avc1.42E01E, mp4a.40.2"'>
</video>
</body>
```

```
</html>< /FONT>
```

Listing 10-21: *Using <video> tag*

8. Save file as "html5_demo.htm". Save it in the same folder where you downloaded the mp4 video. If you want to run the video directly from the Internet, change the **src** attribute to the following path:

http://walisystemsinc.com/sharepoint/art/html5/HTML5_Video_Player_Demo_File.mp4

This html file can be downloaded (in zip format) from the book's website.

If this is your first HTML5 page, then congratulations! You have just created your first HTML5 page. Open the page in IE 9 browser. If you see an image, that means your IE 9 supports the mp4 video format but there is great chance that you will see a red cross on a black background as shown in the image below:

Figure 10-12: *mp4 video format not supported in IE*

To further verify that the video format is not supported in IE, open the following URL in your IE browser:

http://ie.microsoft.com/testdrive/graphics/videoformatsupport/default.html

It shows which video formats are currently being supported in your browser:

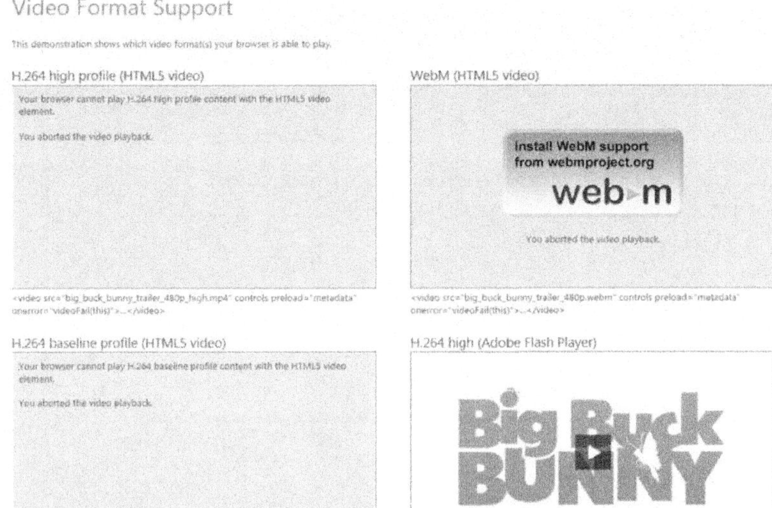

Figure 10-13: *Video formats supported by IE browser*

You will notice that the only support that you have is for Adobe Flash Player. HTML5 video is not supported. Now, what will you do? So next task is to fix the problem and make sure video format is supported by your browser.

So, while your html page is still open in browser, press **F12**. This will open the developer tools.

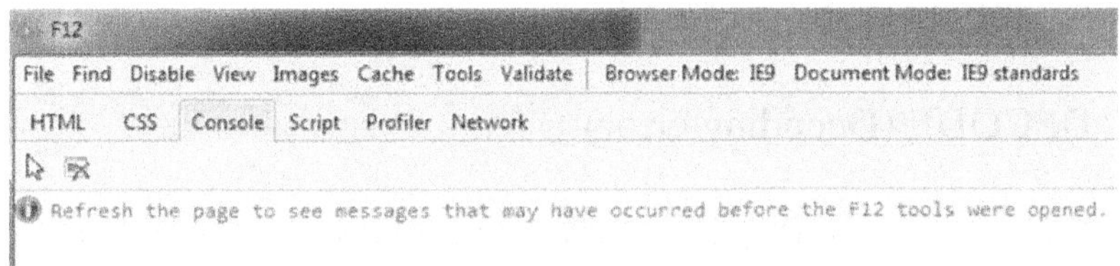

Figure 10-14: *The developer tools in IE 9*

9. Open **Console** tab and enter following line at the bottom and hit **Enter**.

document.getElementsByTagName("video")[0].error.code

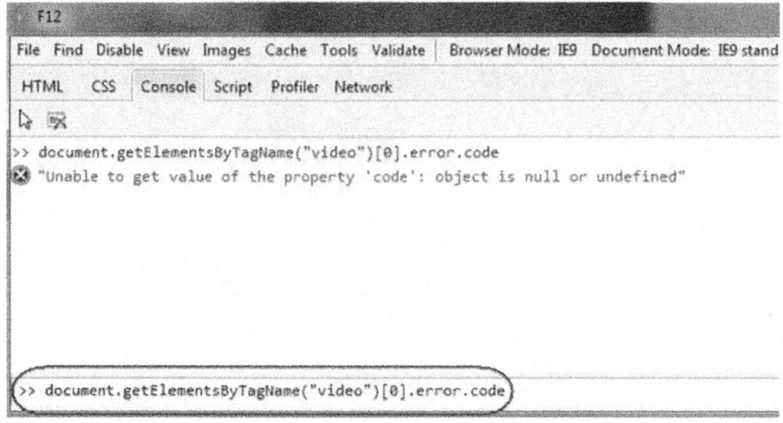

Figure 10-15: *Console Tab*

If there is no error and page loaded correctly, you will get a message:

Unable to get value of the property 'code': object is null or undefined.

This is not an error actually. The page loaded correctly and therefore the error object was null. If there was an error, you would see one of the following results:

1: MEDIA_ERR_ABORTED (Aborted by the user)

2: MEDIA_ERR_NETWORK (Network error)

3: MEDIA_ERR_DECODE (Decoding error)

4: MEDIA_ERR_SRC_NOT_SUPPORTED (Format was not supported)

The first three are temporary issues and can be fixed. Fourth one requires attention. It occurs if encoding is not correct or MIME type is not supported. We have already added the MIME type but if you want to be sure, you can check it in the developer tools.

10. Open **Network** tab in developer tools and click **Start capturing** button.

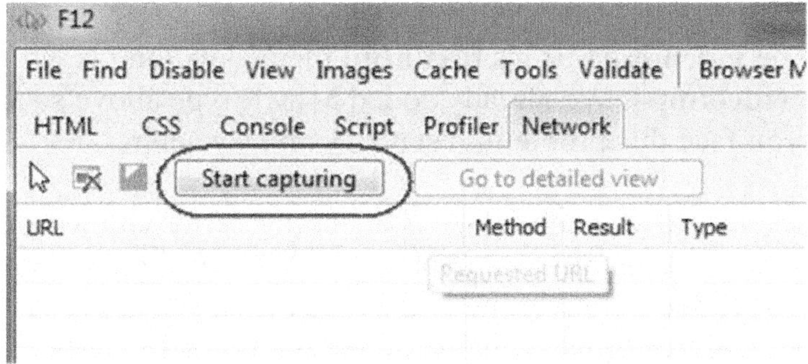

Figure 10-16: *Start capturing*

11. Refresh the html page. Once files have been loaded, you can stop capturing by pressing the **Stop capturing** button.

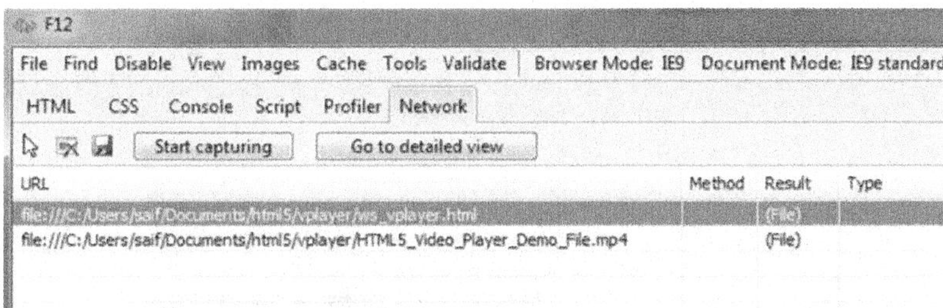

Figure 10-17: *Captured files*

12. Select the video file and click **Go to detailed view** button.

13. Open **Response headers** tab.

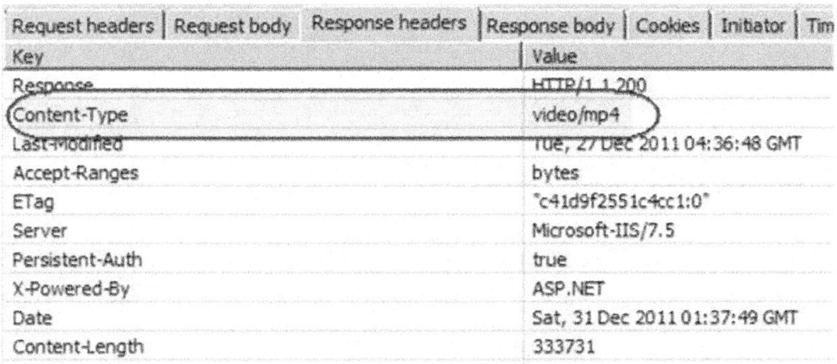

Request headers | Request body | Response headers | Response body | Cookies | Initiator | Tim

Key	Value
Response	HTTP/1.1 200
Content-Type	video/mp4
Last-Modified	Tue, 27 Dec 2011 04:36:48 GMT
Accept-Ranges	bytes
ETag	"c41d9f2551c4cc1:0"
Server	Microsoft-IIS/7.5
Persistent-Auth	true
X-Powered-By	ASP.NET
Date	Sat, 31 Dec 2011 01:37:49 GMT
Content-Length	333731

Figure 10-18: *Content type*

Content-Type should be **video/mp4**. If it's **text/html** then something is wrong. The MIME type should be **video/mp4**. We already added MIME type above so it should work for you. In case you face this problem, restart IIS and try again.

14. Second problem that you could face is encoding. Your video should be properly encoded. As mentioned above already, IE 9 supports only mpeg4 (.mp4) format. Converting and encoding videos is not an issue. There are several free tools available on the net that will do the job.I will discuss two of them below. Of course, you can search and download others also.

15. First one is Movavi Video Converter. You can download 30 days trial version from this location:

http://www.movavi.com/videoconverter!s3941/

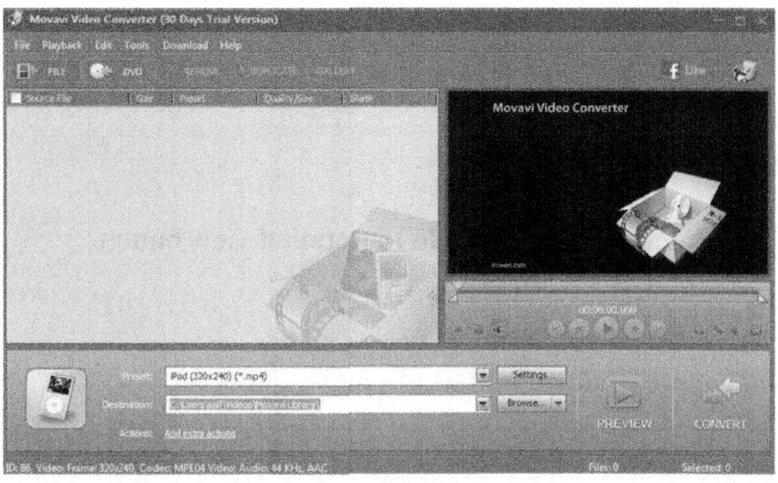

Figure 10-19: *Movavi Video Convertor*

Second one is Miro Video Convertor. You can download it from the following location:

http://www.mirovideoconverter.com/

It's 100% free and open source. Once downloaded, you can open it from **All Programs > Participatory Culture Foundation > Miro Video Converter > Miro Video Converter**.

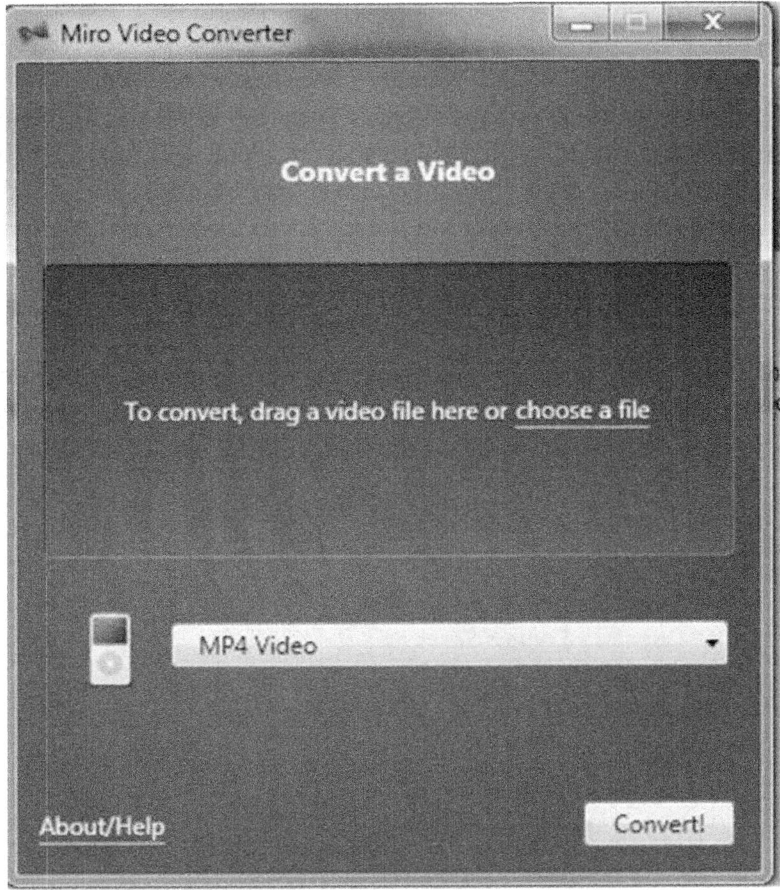

Figure 10-20: *Miro Video Converter*

To convert file to MP4, drag it to the Miro converter and select MP4 from the drop down box. Click **Convert!**. Voila! Your file is now ready to be used in HTML5. It's as simple as that. So, what's next? Do you have necessary codecs on your system? Without required codecs video will still not play even if the file has correct encoding.

Install Video Codecs

It is important that you install the necessary codecs on the machine where you want to run the video. Easiest way is to install Windows media player. Yes, Windows media player is not available out of the box in Windows server 2008 as in other operating systems. It comes installed with Windows 7 but in Windows Server 2008, you have to add it manually. With Windows media player necessary codecs are also installed.

16. There are two ways to install Windows media player. One is to download and install **Windows Media Services 2008 for Windows Server 2008 R2**. By installing these services, you will not get Windows media player but it will give you an option to download and install it on your machine. You can download these services from the following link:

http://www.microsoft.com/download/en/details.aspx?id=20424

17. After you have downloaded and installed these services, open **Server Manager**. Expand **Roles** node. Expand **Streaming Media Services** node. Expand **Windows Media Services** node. Click your server name.

Figure 10-21: *Windows Media Services*

18. On the right side, select **Getting Started** tab. Under **Test your server**, you will see an option to download Window media player. Do not click the link, instead click the green arrow. It will take you to the Microsoft site where you can download Windows media player for your version of Windows. Download the version that is right for your Windows and install it. If you are lucky you will be able to download it and install it. There are reports that there is a problem with this approach. The download page shows Windows media player for XP (64 bit) when you select Windows Server 2008 and after you have downloaded it, it fails to install. If this happens with you, try the second approach that is described below.

In **Server Manager**, right-click **Features** and select **Add Features**. Check **Desktop Experience** box and click **Install**.

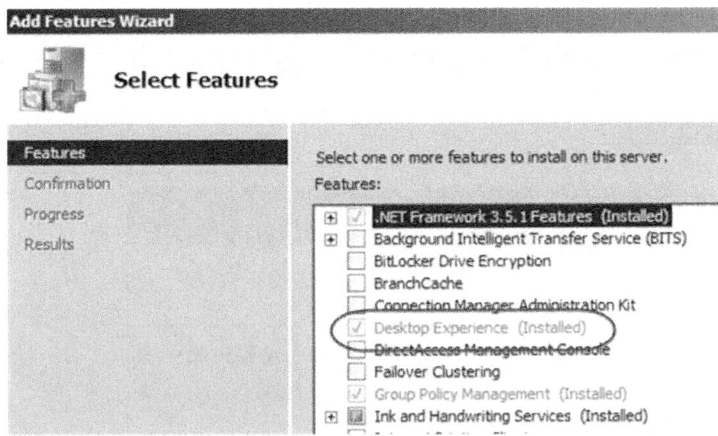

Figure 10-22: *Desktop Experience*

Windows may prompt you for a restart. Restart your machine and now Windows media player will be available on your machine. Try to run the video file directly in Windows media player, if it runs then your machine has the required codecs. You may still get an error message that codecs required to run the file are not available. Don't worry! There is a solution for every problem. Download and install Divx player. That will certainly install all the video codecs on your machine. You can also download codecs separately without installing the player. There are several sites that allow you to download the codecs but downloading the Divx player is the easiest way of getting the codecs. You can download Divx for free from the following site:

http://www.divx.com/

After installing Divx you can play your videos either in Windows media player or Divx player.

By the way, did I mention that IE 9 also supports **WebM** format now? Following is the description of **WebM** (taken from Wikipedia):

WebM *is an audio-video format designed to provide a royalty-free, open video compression format for use with HTML5 video. The project's development is sponsored by Google.*

WebM is a good option for two reasons. First, H.264 is not free. Second, H.264 is a complicated technology. If for some reason, you are unable to run mpeg4 video, you can always try WebM. H.264 is free if your videos are offered for free. If you are creating videos for commercial purposes, then you may have to pay a license fee.

You can download and install WebM support for IE 9 from the following link:

http://www.webmproject.org/ie/

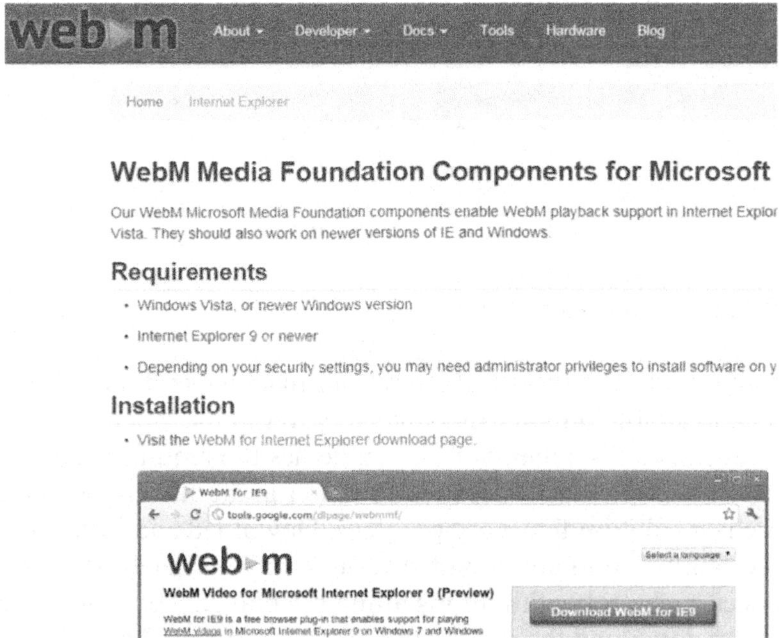

Figure 10-23: *Download and install WebM support*

If your videos are in non-WebM format, use Miro Video Converter to convert them into WebM format.

Alright, so now your system is ready to run the videos. Next step is to write a video player in HTML5.

Building HTML5 Custom Video Player

With the help of Javascript programming, you can create your own controls to control the video. This means you can create your own video player, apply different styles to it, add the functionality you want your way. **Controls** attribute in the **video** tag allows you to use the default controls in the video player. If you remove this attribute, all you get is a video window (without any controls). You can add your own controls underneath the video window.

1. Open notepad and add following code to it.

```
<!DOCTYPE html>
<html>
<body>
<video id="video1" controls>
<source src="HTML5_Video_Player_Demo_File.mp4" type='video/mp4;
codecs="avc1.42E01E, mp4a.40.2"'>
</video>
</body>
</html>
```

Listing 10-22: *Layout controls*

2. Save the file as **vplayer.html**. Now we are going to make changes to this file and add new code.

3. Add following CSS code between the <head> and </head> tags.

```
<STYLE type="text/css">
#bar
{
border: solid 1px black;
width: 100%;
margin-bottom: 5px
}

#position
{
height: 15px;
color: white;
background: steelblue;

}

#mask
```

```
{
float: left;
position: relative;
padding: 10px;
border: 5px solid #61625d;
font-family: Arial, Helvetica, sans-serif;
background: #000000;
box-shadow: inset 0 15px 35px #535353;
}

#vcontrols
{
position: relative;
padding-top: 10px;
float: left;
clear: both;
width: 98.5%;

}
section {
display: block;
}

.boldbuttons{
display: block;
float: left;
font: bold 13px Arial; /* Change 13px as desired */
line-height: 22px;
height: 30px;
padding-left: 8px;
text-decoration: none;

}


.buttonspan{
display: block;
padding: 1px 8px 1px 8px;
}

.buttonwrapper{
```

```
overflow: hidden;
width: 100%;
}

</STYLE>
```

Listing 10-23: *Video player style sheet*

This CSS code will be used to stylize the video player. You can write your own style sheet to add more style to the video player and the controls. You can also create different skins and apply them dynamically to the player. Discussing that is beyond the scope of this section. Our goal is to run HTML5 code in SharePoint 2013.

4. Replace the following code with new code (shown below in code listing):

```
<video id="video1" controls>
<source src="HTML5_Video_Player_Demo_File.mp4" type='video/mp4;
codecs="avc1.42E01E, mp4a.40.2"'>
</video>
```

```
<section>
<div id="mask">
<video id="video1" ontimeupdate="progressBar()">

<source src="HTML5_Video_Player_Demo_File.mp4" type='video/mp4;
codecs="avc1.42E01E, mp4a.40.2"'>
</video>

<div id="bar">
<div id="position" >
</div>
</div>

<div id="vcontrols">

<button onclick="play(); return false;" class="boldbuttons"><span
class="buttonspan">Play</span></button>
<button onclick="pause(); return false;" class="boldbuttons"><span
class="buttonspan">Pause</span></button>
```

```
<button onclick="stop(); return false;" class="boldbuttons"><span
class="buttonspan">Stop</span></button>
<button onclick="fast(); return false;" class="boldbuttons"><span
class="buttonspan">Fast</span></button>
<button onclick="slow(); return false;" class="boldbuttons"><span
class="buttonspan">Slow</span></button>

</div>

</div>
</section>
```

Listing 10-24: *Controls layout*

Code Explanation

You have given an ID to the video tag. This ID is used in the Javascript code. We have added five buttons under the video tag. These buttons are as following:

- **Play**: To play the video
- **Pause**: To pause the video. Video resumes from the same place where it was stopped when played again.
- **Stop**: To stop the video. Video starts from the start when played again.
- **Fast**: To increase the speed of the video. Video returns to normal speed when **Play** button is clicked.
- **Slow**: To reduce the speed of the video. Video returns to normal speed when **Play** button is clicked.

Each button has a function attached to it. Function is called when **click** event is fired. Note there is a slight change of behavior in SharePoint 2013 web parts. When you click button to call a javascript function, the whole page refreshes which interferes with code functionality. In our case, the player fails to run because button click refreshes the page and the player is back to the original state. This did not occur in SharePoint 2010. To resolve it, add **return false;** in the **onClick** event of the button, after calling the javascript function. That will stop thc page postback.

The outer <div> (<div id="bar"> is used to draw a solid border. The inner <div> (<div id="position">) is used to indicate the current playback position. When video playback is under way, the <video> tag triggers the **onTimeUpdate** event continuously. You call **progressBar()** function to update the progress bar.

5. Javascipt functions are given below. Copy them where section ends (after **</section>**tag).

```
<script type="text/javascript">

function progressBar()
{
var position = document.getElementById("position");
var video=document.getElementById("video1");

var status=document.getElementById("progress");

position.style.width = (video.currentTime/video.duration * 100) + "%";

//status.innerHTML = (Math.round(video.currentTime*100)/100) + "
sec";

}

function play()
{

var video=document.getElementById("video1");
video.play();

}

function pause()
{
var video=document.getElementById("video1");
video.pause();
}

function stop()
{
var video=document.getElementById("video1");
```

```
video.pause();
video.currentTime = 0;
}

function fast()
{
var video=document.getElementById("video1");
video.play();
video.playbackRate = 12;
}

function slow()
{
var video=document.getElementById("video1");
video.play();
video.playbackRate = 0.25;
}

function normal()
{
var video=document.getElementById("video1");
video.play();
video.playbackRate = 1;
}

</script>
```

Listing 10-25: *Javacript adds functionality to the player*

Code is pretty straight forward and self-explanatory. **progressBar()** function is used to create the progress bar. **normal()** function is not being used as it served the same purpose as offered by the **Play** button. You use **document.getElementById()** to get hold of the video tag.

Save the file again.

You can download the video used in this exercise from the book's website:

Copy the video file to the same folder that contains the html file. If you want to use your own video, just change the video path in the **src** attribute in the code.

Using HTML5 in SharePoint 2013

Finally, time to take the HTML5 code to SharePoint, the step you were waiting for so anxiously!

6. There are two ways to write HTML5 code in SharePoint. First is to use a content editor web part and second is to write the code directly inside a SharePoint page. We will cover both approaches below. First let's try the content editor approach.

7. Open your SharePoint site in browser. Create a new page to showcase the web part. Click **EDIT** link in the top right corner to edit the newly created page. Select **INSERT** tab from the top. Click **Web Part** button in the ribbon. In the **Categories** section, select **Media and Content**. In the **Parts** section, select **Content Editor** and click **Add** button.

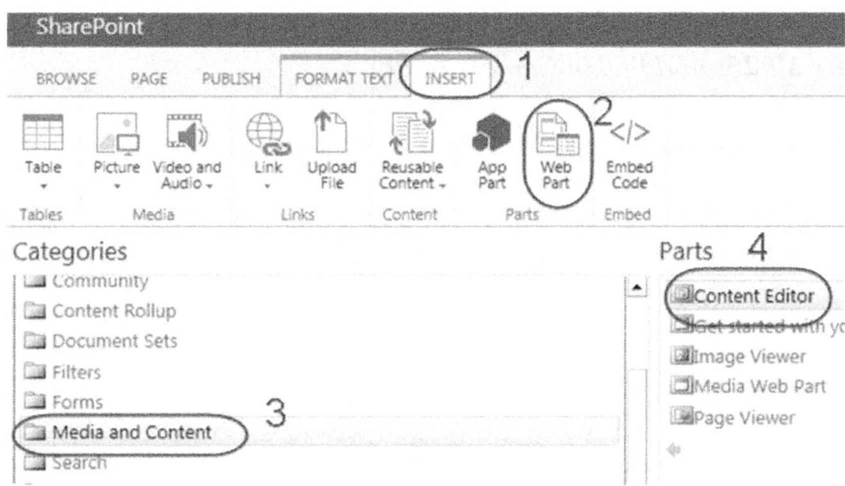

Figure 10-24: *Add web part*

8. Now that you have added the **Content Editor** web part, there are two ways to add HTML5 code. One is to create a site in IIS to host the html page and then provide the site link in the **Content Editor** web part. Second is to add the HTML5 code directly in the web part. Let's explore both options below.

9. Let's try the IIS option first. Create a site in IIS and point it to the folder that contains the html file. Test the link in the browser. Now, edit the web part and add the link in the **Content Link** box. Click **OK** to apply the change.

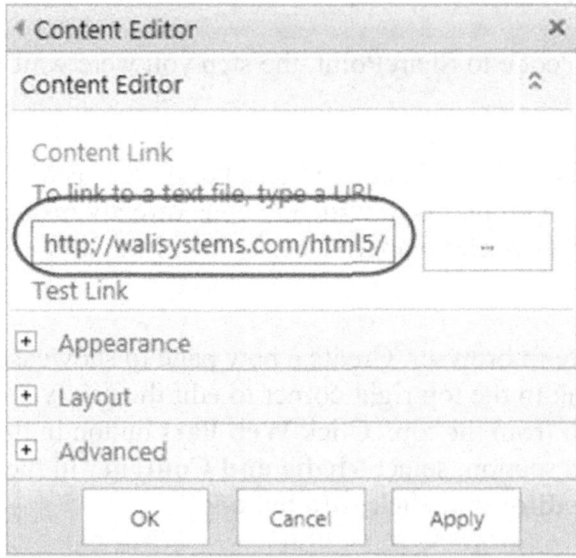

Figure 10-25: *Add link in Content Editor*

10. Click **Save** button (in the **PAGE** tab) to save the page.

11. Test the code. Click **Play** button in the player to play the video. Following screenshot shows how the video player looks inside a SharePoint 2013 page.

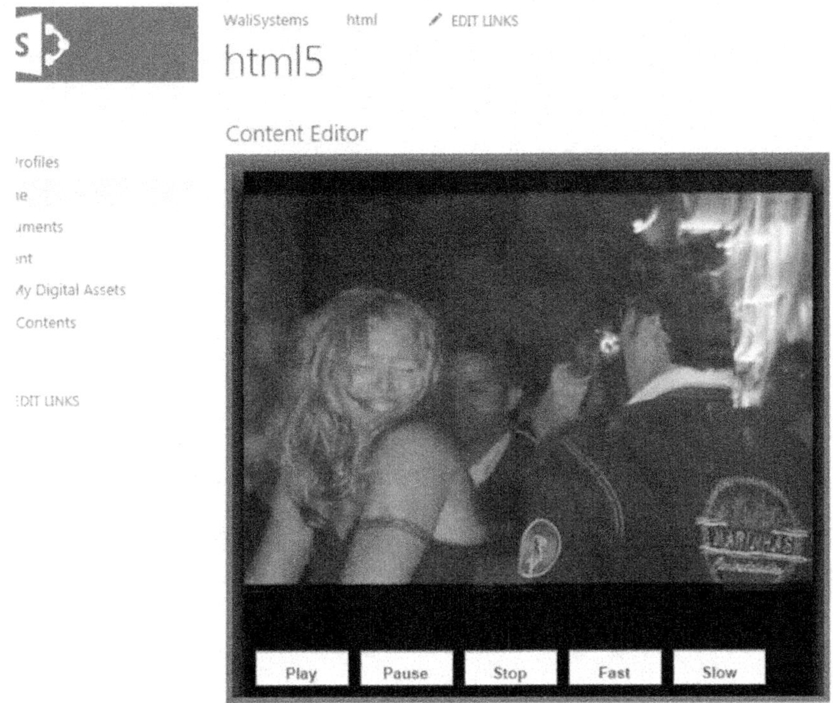

Figure 10-26: *Video Player in Action inside SharePoint 2013 page*

You can also embed this player in script editor. Script editor is a new addition in SharePoint 2013. Content editor is used to display any kind of content on a SharePoint page. You can copy content to the content editor and then use SharePoint editor or HTML editor to edit and format the content. Script editor web part is specifically used to add script to the page. You can put html or javascript code in the web part and add functionality to the page. The process of adding script editor web part and adding code to the web part is same as that of adding content editor web part described above.

This was the first option. Now, let's try the second option.

12. Edit the web part again and remove the link from **Content Link** field. Click **OK**.

13. Again, edit the web part by opening the menu from inside the web part.

14. Click inside the web part and then click **Edit Source** button in the ribbon.

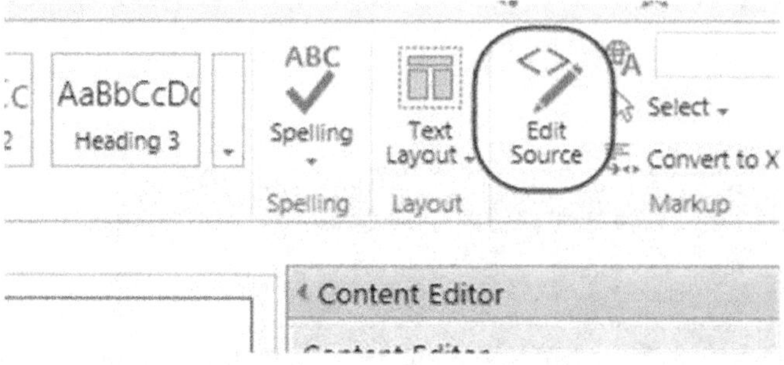

Figure 10-27: *Edit HTML Source*

15. This will open **HTML Source** editor. Open html file in notepad and copy all content and paste it in the **HTML Source** editor and click **OK**.

HTML Source

```
}

#position
{
height: 15px;
color: white;
background: steelblue;

}

#mask
{
float: left;
        position: relative;
        padding: 10px;
        border: 5px solid #61625d;
        font-family: Arial, Helvetica, sans-serif;
        background: #000000;
        box-shadow: inset 0 15px 35px #535353;
}

#vcontrols
{
```

Figure 10-28: *HTML Source editor*

16. Click **OK** in the **Content Editor** to save the changes. Remember if you have used script editor web part then there is **EDIT SNIPPET** link inside the web part that you click to open the source window.

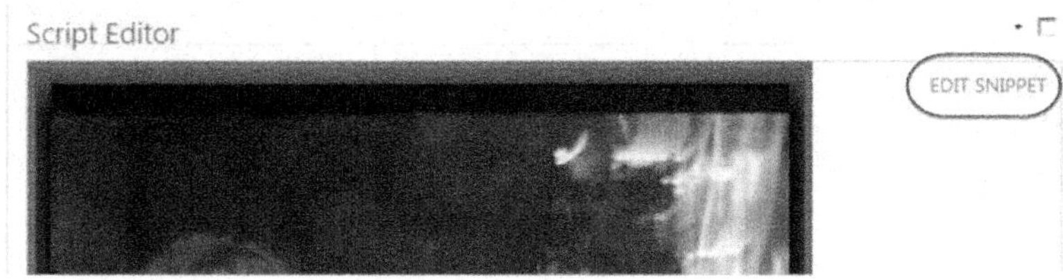

Figure 10-29: *Edit Snippet link in Script Editor*

That's it. Save and publish the page. Click the **Play** button to play the video in the player. You can also use this player in the **Asset Library** and use it to play videos from the library. Since this is custom built player, you can even add features that are missing in the default SharePoint player.

Now, let's implement this player in a new page.

17. Open SharePoint site in SharePoint Designer. Click on **Site Pages** from the **Navigation**.

18. Click **Page** button (click the arrow to open the submenu) in the ribbon and select **ASPX**.

19. Replace **Untitled_1** with **videoplayer** to rename the file. The file name will be videoplayer.aspx.

20. Right-click videoplayer.aspx and select **Check out**. Again right-click and select **Edit File in Advanced Mode**.

21. Locate **<title>** tag and change the title from **Untitled_1** to **HTML5 Video Player**.

22. Add following code before the **</head>** tag.

```
<STYLE type="text/css">
#bar
{
border: solid 1px black;
width: 100%;
margin-bottom: 5px
}
#position
{
```

```
height: 15px;
color: white;
background: steelblue;
}
#mask
{
float: left;
position: relative;
padding: 10px;
border: 5px solid #61625d;
font-family: Arial, Helvetica, sans-serif;
background: #000000;
box-shadow: inset 0 15px 35px #535353;
}
#vcontrols
{
position: relative;
padding-top: 10px;
float: left;
clear: both;
width: 98.5%;
}
section {
        display: block;
}
.boldbuttons{
display: block;
float: left;
font: bold 13px Arial; /* Change 13px as desired */
line-height: 22px;
height: 30px;
padding-left: 8px;
text-decoration: none;
}

.buttonspan{
display: block;
padding: 1px 8px 1px 8px;
}

.buttonwrapper{
```

```
overflow: hidden;
width: 100%;
}
 </STYLE>
```

Listing 10-26: *Style sheet*

23. Remove **<form></form>** tags from the body section and insert following code after **<body>** tag.

```
<section>
<div id="mask">
<video id="video1" ontimeupdate="progressBar()">

<source src="HTML5_Video_Player_Demo_File.mp4"
type='video/mp4; codecs="avc1.42E01E, mp4a.40.2"'>
</video>

<div id="bar">
<div id="position"  >
</div>
</div>

<div id="vcontrols">

<button onclick="play(); return false;" class="boldbuttons"><span
class="buttonspan">Play</span></button>
<button onclick="pause(); return false;" class="boldbuttons"><span
class="buttonspan">Pause</span></button>
<button onclick="stop(); return false;" class="boldbuttons"><span
class="buttonspan">Stop</span></button>
<button onclick="fast(); return false;" class="boldbuttons"><span
class="buttonspan">Fast</span></button>
<button onclick="slow(); return false;" class="boldbuttons"><span
class="buttonspan">Slow</span></button>

</div>
</div>
</section>
```

```
<script type="text/javascript">
function progressBar()
{
var position = document.getElementById("position");
var video=document.getElementById("video1");

var status=document.getElementById("progress");
position.style.width = (video.currentTime/video.duration * 100) + "%";
}

function play()
{
var video=document.getElementById("video1");
video.play();
}

function pause()
{
var video=document.getElementById("video1");
video.pause();
}

function stop()
{
var video=document.getElementById("video1");
video.pause();
video.currentTime = 0;
}

function fast()
{
var video=document.getElementById("video1");
video.play();
video.playbackRate = 12;
}

function slow()
{
var video=document.getElementById("video1");
video.play();
video.playbackRate = 0.25;
```

```
}

function normal()
{
var video=document.getElementById("video1");
video.play();
video.playbackRate = 1;
}

</script>
```

Listing 10-27: *Video player code*

Summary

You are not dependent on Flash anymore. You can create rich Internet applications with HTML5. Not only that, you can now provide same rich functionality with simple code (no clutter, no mess). You can create rich applications for both web and mobile devices. Create animations using Canvas. You can create amazing applications by using HTML5 with Javascript. Geolocation is a cool new feature that lets you find out where in the world your visitors are. You get latitude and longitude coords to find the position of the user. Adding audio and video to your web pages is a breeze now.

Chapter 11

SharePoint Online

In this chapter:

- SharePoint sites in the cloud
- SharePoint Online Packages
- Setting up SharePoint Online

SharePoint in the cloud

You will kill two birds with a stone in this chapter. The objective is to shed light on the very interesting topic of SharePoint branding. Since SharePoint is very popular with the enterprise clients, they mostly like to add their own branding to the product. There was a time when the only thing you could do "easily" was changing the logo and the site's title and description. Adding elements to the page's style was a cumbersome job. When Microsoft saw companies' interest in this area, they started adding some support in the product and made it easier for the users to add branding. Still it was not where it should have been. With SharePoint 2007, things became somewhat easier. With SharePoint 2010, branding actually took off and we saw some great very well done sites. Now with SharePoint 2013, we have full support in the product and branding sites has become a lot easier now. With the increasing popularity of mobile platforms, responsive designs have taken the top spot and are the #1 preference of the companies these days. Moreover, SharePoint is now very well established in the cloud. SharePoint until now has been popular as an Intranet product. Companies usually used it for team collaboration and document management but with SharePoint online, companies are now deploying sites that offer the same excellent look and feel that was the trademark of HTML sites back in the day and on top of that, users get the same excellent features that were the hallmark of SharePoint Intranet or Extranet solutions, for example, document management, users management, user roles, calendars, tasks management, etc. So, in this chapter, first we are going to setup a SharePoint site in the cloud. Then we will develop a branded site for the cloud.

So, here is a brief introduction before we dive into the details. SharePoint Online is a cloud based service that helps organizations share and collaborate with colleagues, partners, and customers. Users can now access internal sites, documents, and other information from anywhere. It is a common platform that can be accessed from the home, office or a mobile device.

SharePoint Online has new innovative features. It has a fluid interface, file sharing capabilities, higher storage limits, cloud application development model and robust tools for administrators to more easily control the environment. Everyone benefits from the new SharePoint Online: users, managers, developers and IT professionals.

SharePoint Online site consists of two sites: the Team Site and the Website. Here is some more detail:

Team Site: The Team Site is an internal site for your team. All members of your organization can collaborate with each. They can share documents and other files. Other basis elements of the Team site include announcements, meetings, tasks list and issues list, custom lists, etc. You can create subsites from the available site templates. For example, you can create a team blog, or a social meeting workspace to organize team events, meetings, and so on.

Website: The Website is where you design your public site. This is the site that is visible to the world. This website has everything you need to get started with designing your site. It has libraries to store content such as images, documents, etc. that you can display on the pages of your site.

Site Components

A site is a group of pages where users can work on projects, store and share information. Following are some important site objects that are frequently used in a SharePoint site.

Lists

A list is a site component that is used to store information quickly. It has CRUD functions. There are different types of lists in SharePoint, for example, a tasks list is used to store and monitor tasks, an events list is used to store events, a custom list can comprise of any number of fields that are relevant to your business.

Libraries

A library is a special type of list that stores files as well as information about files. The two most used libraries in SharePoint are: Document library and Forms library.

As the name suggests, document library is used to store documents. You can create folders inside a library to define a hierarchy. Forms library is used to store electronic forms. You can store forms directly in this library. Each form library can be associated with different content types. You can publish InfoPath form template in a library and bind it to the library. Users can open this template and fill out data. Saving the template automatically generates a new xml form that is stored in the same library.

Views

Views play an important part in SharePoint. You can use views to see items in a list or library. These views can be added to the home page of your site so that all important information from different libraries and lists is available on the home page of the site. Multiple views can be created for a list or library. You can also use a web part to display a view of a list or library on a separate page of the site.

Web Parts

A Web Part is a modular unit of content that forms a basic building block of most pages on a site. If you have permission to edit pages, you can use web parts to customize the site to display information from different sources.
New interface has been reimagined. It is easy to navigate and easy to accomplish tasks.

Site Usage

There are several other features that allow administrators to control the site. Managing sites in SharePoint is easy because of the built-in functions that let administrators control permissions, roles, storage, content management, etc.

Permissions

There are different permission levels available out-of-the box and you can create your own custom permission levels. You can assign permissions directly to the users as well groups. Assigning permissions to the groups make permission handling very easy. All you have to do is add or delete users from the group. You don't have to handle permissions individually if you don't want to. For granularity, you can assign permissions directly on the files and folders and to the individual users.

Customization

Everything can be customized in SharePoint and that's the beauty of it. You can customize the interface, you can customize permissions, you can customize navigation. This is the biggest advantage of SharePoint. You are not restricted to the out-of-the-box features. SharePoint is scalable. You can add your own custom functionality.

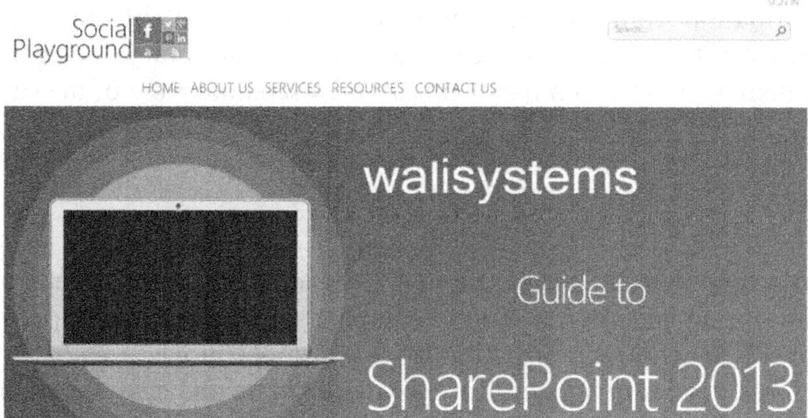

Figure 11-1: *SharePoint 365 (branded) home page*

The page you see above is customized and contains custom content. One new feature is drag and drop. Another one is on-hover that allows users to see a set of rich commands and contextual information in search results.

SkyDrive Pro is cloud storage that users can use to store documents directly from within SharePoint.

Most important interesting new feature is new and improved public website. There are different out-of-the-box themes available that can quickly change the look and feel of your site. You can also create custom themes. Even if you don't create custom theme, you can create a new design and create a master page to incorporate the new design and apply it to your site. Following are a couple of themes that come with the product.

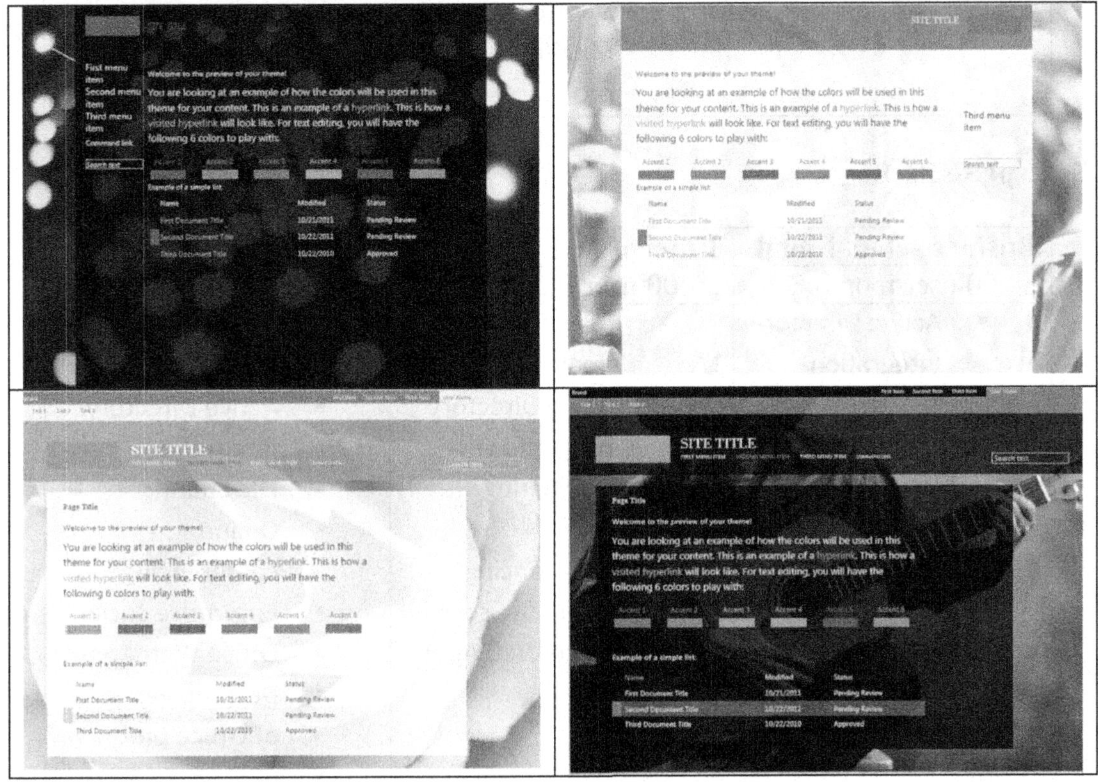

Figure 11-2: *Out-of-the-box themes*

Online Packages

As of Dec 2013, the packages are as following:

Small Business Packages

Small Business	**Small Business Premium**	**Midsize Business**
$5.00 user/month	$12.50 user/month	$15.00 user/month
Business-class email ▪ Online conferencing ▪ Public website ▪ File sharing ▪ Office Web Apps	▪ Business-class email ▪ Online conferencing ▪ Public website ▪ File sharing	▪ Business-class email ▪ Online conferencing ▪ Public website ▪ File sharing

	▪ Office Web Apps ▪ Desktop versions of Office products	▪ Office Web Apps ▪ Active Directory Integration

Enterprise Packages

Business-class Email	Enterprise E1	Enterprise E3
$4.00 user/month	$8.00 user/month	$20.00 user/month
▪ Active Directory Integration	▪ Business-class email ▪ Active Directory Integration ▪ File Sharing ▪ Online conferencing ▪ Enterprise social ▪ Office Web Apps	▪ Business-class email ▪ Active Directory Integration ▪ File Sharing ▪ Online conferencing ▪ Enterprise social ▪ Office Web Apps ▪ Desktop versions of Office products ▪ eDiscovery Center

Setting up SharePoint Online

In this section, we will start from scratch and setup an online SharePoint site.

Sign-up for a trial version on this link:

http://office.microsoft.com/en-us/business/compare-office-365-for-business-plans-FX102918419.aspx#

Click **Free trial** link under Enterprise E3 plan and sign up.

Here is the direct link:

https://portal.microsoftonline.com/Signup/MainSignUp15.aspx?OfferId=B07A1127-DE83-4a6d-9F85-

2C104BDAE8B4&dl=ENTERPRISEPACK&Country=US&culture=en-us&WT%2Eintid1=ODC%5FENUS%5FFX102918419%5FXT103040305&ali=1

Trial version expires after 30 days. If you choose to keep the site, then you have to purchase the package.

After you have signed up, you get access to the site. You get your own URL. You can also unique URL for your site should you wish to do so. You can purchase a domain name and link it to your SharePoint Online site.

Site URL looks like this:

http://sitename-public.sharepoint.com

You are also granted a unique email address that you use to login to your site. Your email address looks like this:

username@sitename.onmicrosoft.com

Plan the Public Website

So, now you have a subscription, you can setup a public website and make it available on the Internet. By default, your site is set to Offline mode. Once you have completed the site, you can take in Online. You can customize the website to create a professional online presence for your business or organization.

It's always a good idea to do some planning before you start the work. Discipline and planning really help to keep things moving smoothly.

The website includes several sample pages: Home, About Us, Contact Us, Directions, and Blog. You can use these pages and just customize the content and you are all set to go live. You can add your company name, logo, contact information, and change color scheme. You can also create a totally new branded site for your business. That's our next topic in this chapter. In this section, we will use out-of-the-box components to set up a site.

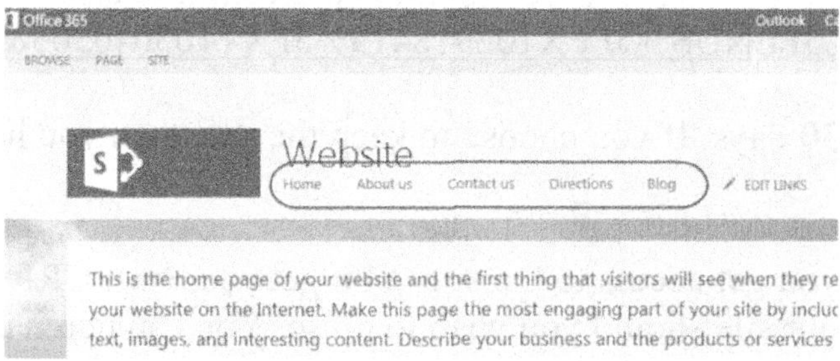

Figure 11-3: *Website menu*

Once you login, you will see a global menu. As mentioned before, there are two sites created by default, a team site and a public website. To work on the team site, you have to select team site from the global menu. To work on the public site, you have to select public website from the menu. Global menu looks like this:

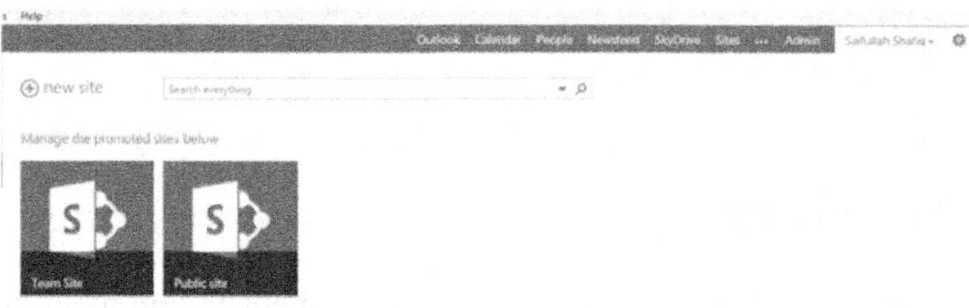

Figure 11-4: *Global menu*

In the global menu, click **Sites** to see the two sites. Please note that depending on your package, the interface may look different. For example, if you are using an E3 package, you will not see the two sites. You will have to create the sites first. By default, E3 package gives you a public site. You can create more sites separately. Click **Public** site to go to the public website. Click on the **Home** link to go to the home page. Then select **PAGE** tab from the top. Click **Edit** button from the **PAGE** menu. Start typing and adding elements to the page until you are done. If you have already designed a page using other design programs, copy the page from there and paste it in the content area and the save and publish the page.

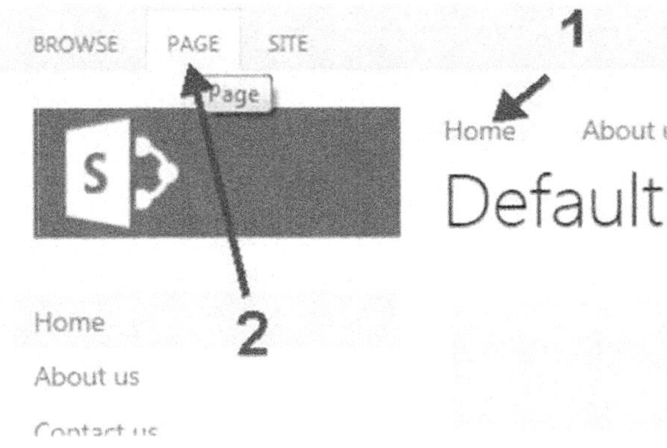

Figure 11-5: *Page tab*

Repeat the process for all other links. Once you have added content to the pages, it's time to make a few more changes to the site. Go to the **SITE** tab.

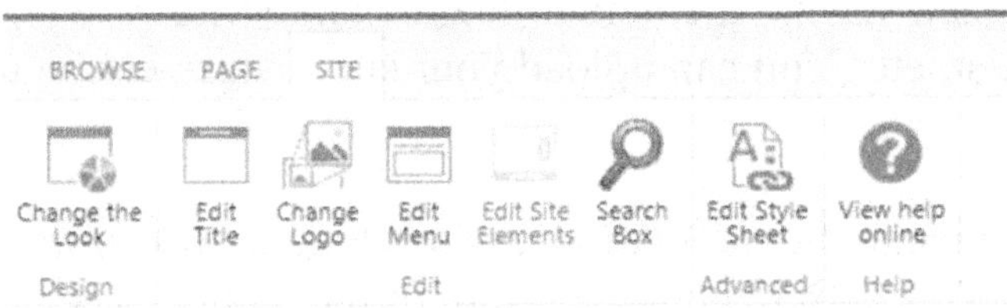

Figure 11-6: *Site tab*

Click **Change the Look** button. This page lists all out-of-the-box themes. If you are ok with the default theme, you don't have to do anything but If you want to change your theme then this is the place. Just select the theme you like and your site's look and feel will change.

Click **Edit Title** button to change the title of the site. Add title and description and click **Ok**.

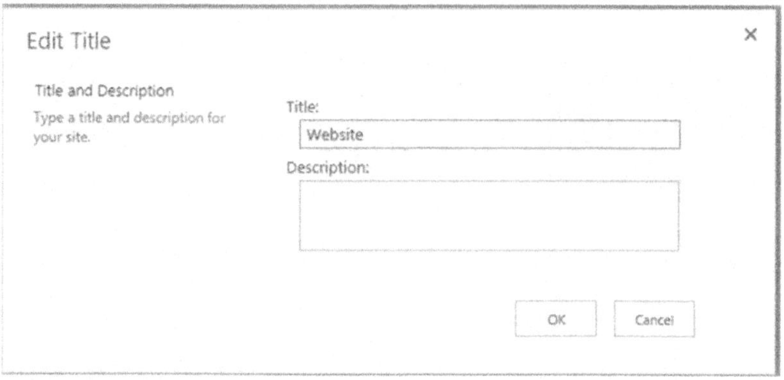

Figure 11-7: *Site Title*

Click **Change Logo** button to change logo on your site. You can upload logo image from your computer or from a SharePoint site if you have already upload an image to your site. Each public website has all the default features like Site Assets, Document Library, Tasks List, etc. You can upload your site images etc. to the **Site Assets** library.

Click **Edit Menu** button to change the navigation menu. Menu is shown on the left side. There is a link called **EDIT LINKS** at the bottom of the menu list. Click it to turn menu into editable mode. Click the link you wish to edit. You can hide or unhide links from the menu. You can also drag new links to the menu or use the new **link** to add a link to the menu. After making the changes, click the **Save** button.

Next, click **Search Box** button if you want to hide/unhide Search box on your site.

If you want to make changes to the design of your site, you can edit its style sheet by clicking the **Edit Style Sheet** button. You can either type styles directly to the input box or you can copy styles from the style sheet file you may have created for your site. Click **Save and Publish** to save and publish the styles. If you clicked **Save** button, then that will save the styles but won't publish it and thus changes will not appear on your site. To enable the changes, make sure to publish the style sheet.

Now, one final thing, you can edit properties of each page in your site and set its SEO properties as well. Go to the home page and click **PAGE** tab. **Edit Properties** button has three options: Edit Properties, Edit SEO Properties, View Properties. Select **Edit SEO Properties**. This form has all the properties that are needed to setup SEO for your page. If you want to rename your page, do it in the **Name** field. Add title of the page in the **Title** field. This title is used by the search engines to pair the page with results for similar keywords so make sure you had all keywords in the title field that are relevant to your business, for example, "Wali Systems – A

SharePoint Company – SharePoint Staffing, Consulting and Development Services". **Browser Title** is the title that appears in the title bar of your page. **Meta Description** contains the description of your page. This is displayed under your URL shown in search results. **Keywords** field is the field where you enter the keywords for your site. Keywords should be comma separated. You can enter maximum of 1024 characters in this field. In **Exclude from Internet Search Engines field**, you can choose if you want to include this page in the sitemap and expose to the search engines or you want to keep it hidden. After entering values for each field, click OK to save the changes. Repeat this process for all the pages.

Now your site is ready to be published but before doing that let's explore the admin section as well. Click **Admin** link from the top global navigation.

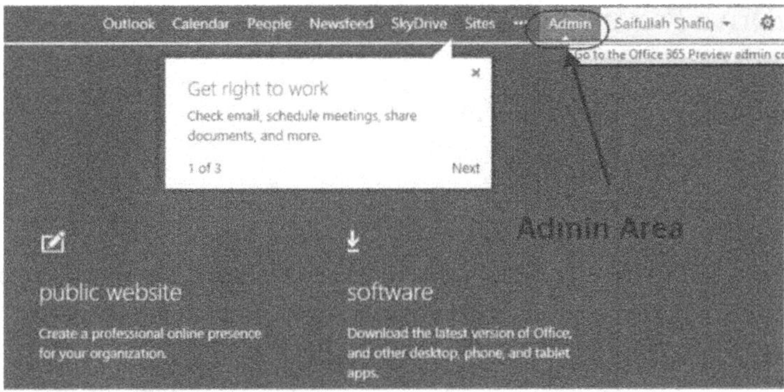

Figure 11-8: *Admin area*

Here you get all the options to make necessary changes to your site. For example, by default your site's email is set to **username@sitename.onmicrosoft.com**. You can change it **username@yourcompany.com**. To change email address, click the box labeled **email address**. Changing email process takes 45 minutes. You must have a domain name before you start this process. Click the **Start** button to start the process and enter you email with your domain name and click **Next**. On the next screen, confirm that you own the domain name. Provide name and password for your domain registrar. This is where you bought your domain. Enter username and password for your DNS hosting provider. This may be the same company as your domain registrar. To confirm that you own the domain name, you have to login to your domain account and add a **DNS** entry. In the **DNS** records, add a **TXT** entry. Enter the code provided by Office 365 site. This is the first step. In the next step, you confirm that you want to keep your website where it's hosted today and just want the email address to be associated with your SharePoint site. In the third step, you change your email address. Enter new email to link it to your current SharePoint email and click **Update**. In step four, you add other users of your site that have a custom email address. Remember you have to setup these users in

SharePoint also using their custom email addresses. In the final step five, you redirect all traffic from your site to Office 365 site. To make this change, you have to change your name servers in the domain account. Name servers will be provided by the Office 365 site, you just have to enter them in your domain account. Once you have completed all the steps, you will start getting emails at your custom email address associated with your SharePoint website.

Click **collaboration** button to see how to use tools for creating, sharing, and tracking documents. You do not configure anything in this section. This section serves as a small tutorial for the features that you can use in your site. Here is what you will learn in this section:

- Check out SkyDrive Pro and team sites
- Get to know SkyDrive Pro
- Change the look and feel of your team site
- Put your site to work for the team

Click **public website** box/button to learn more about creating public website. Click software button to setup latest software for your site. This section records the computer name where you download and install the software. It records the operating system and the installation date. Once installed, you can deactivate the installed software using the **deactivate** link at a later time. You can setup **Lync** in this section. You can setup **tools and add-ins**, for example, SharePoint Designer. Remember you can use these tools from your desktop as well if you have them installed already. You just need to connect to your online site from the desktop tool. More on this topic later! You can setup desktop applications if you have Office 2007 or 2010 installed. If you have Office 2013 installed, you don't have to do anything. You can setup Office 365 on your mobile devices as well. Remember you can use your mobile's browser to access your Office 365 site but for a better browsing experience or if you want to access email on your mobile device, or you want to setup OWA for your mobile device, you can set that up in this section.

Choose your phone or tablet

Phone	Tablet
Windows Phone	Windows 8 tablet
iPhone	Windows RT device
Android Phone	iPad
BlackBerry®	
Nokia (Symbian OS)	
Other	

Figure 11-9: *Supported mobile devices*

All devices listed above in the figure are supported. As an example, we will set it up for iPhone in this section. In the **software** section, click **phone & tablet** link on the left side then click on **iPhone**. Click **get apps** button. It will take you to another page. Enter your mobile number and click **Submit** to receive a link in the form of a SMS or you can directly open the link www.office.com/business-apps on your mobile device. The mobile apps shown below are available on your iPhone:

- Word
- Excel
- PowerPoint
- OneNote
- OWA
- Lync
- SharePoint

You see following page when you open the above link. Click on **OWA** to go to the App Store and install the application. It's free.

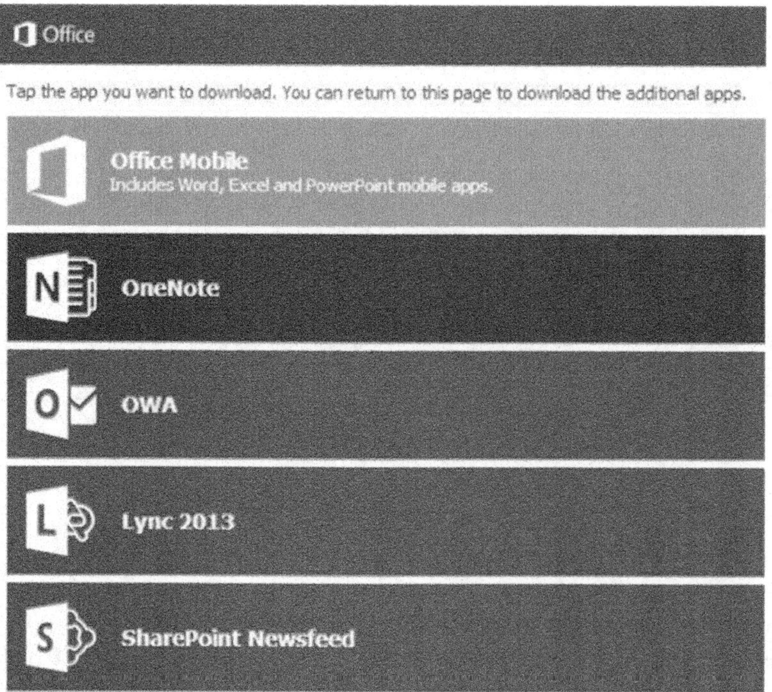

Figure 11-10: *Office Web Apps*

After installing, click **Open** to open the application. Click **continue** and login with your username and password (for your SharePoint online site). It may take a few seconds to find your server. After a few seconds, you will see a welcome screen. You will be prompted to allow OWA to access your contacts. Click **Ok**.

Once signed in, you can switch between Mail, Calendar, People and Options by tapping on the icon in the bottom left corner of the app.

to get around

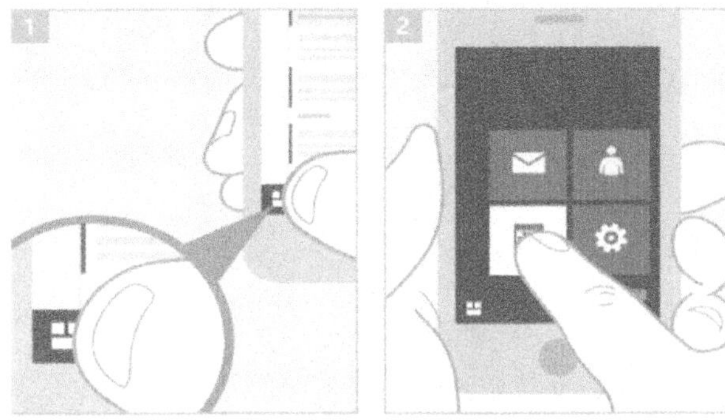

To switch between Mail, Calendar, People, and Options, tap the icon in the bottom left corner.

Figure 11-11: *Navigation*

This takes you to a navigation screen with titles you can tap to reach the desired module.

Figure 11-12: *Say it*

You can tell your device what you want it to do. Tap and hold the "go to" button and say what you want the application to do. This is a great feature but isn't available in all languages.

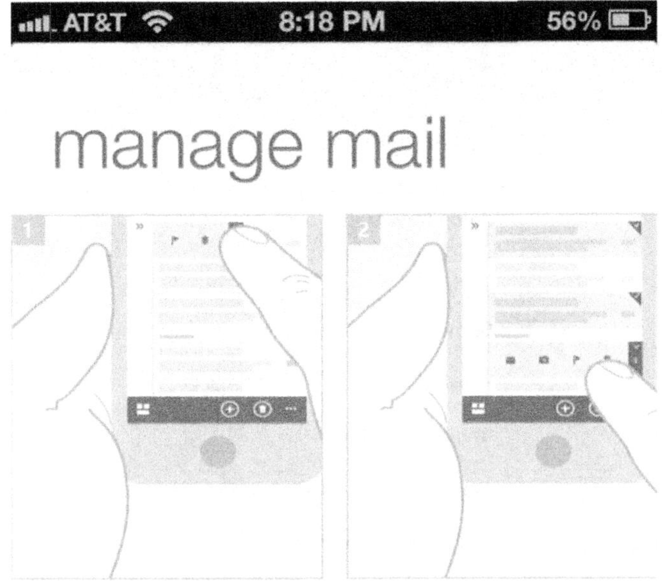

Figure 11-13: *Manage mail*

A quick swipe of each list view item reveals a set of actions such as delete, flag, move and mark as read/unread. In the **Inbox**, you see all messages. In **Calendar**, you can add new events. You can easily swipe from day-to-day, week-to-week or month-to-month, and you can tap on any calendar item to display a preview of the event details. In the **People**, you see all contacts. Your OWA contacts are automatically synced to the native address book on the device.

You can also setup Office 365 email on your iPhone using SMTP settings. The server name for IMAP and POP is outlook.office365.com and the server name for SMTP is smtp.office365.com. These settings can be used if you are using the latest version of Office 365.

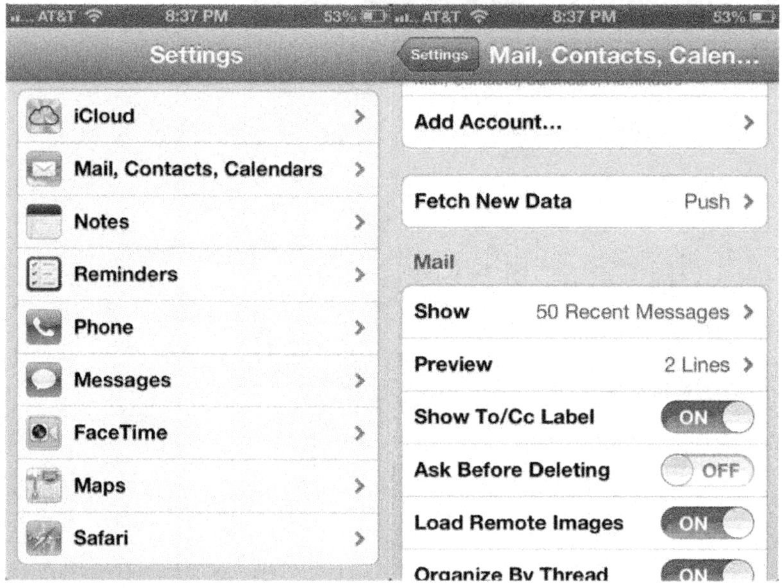

Figure 11-14: *Email settings in iPhone*

Go to iPhone settings and select **Mail, Contacts, and Calendars**. Select **Add Account**. Select **Other**. Select **Add Mail Account**. Add **Name, Email, and Password** and hit **Next**. Under **Incoming Mail Server**, enter outlook.office365.com in the **Host Name** field. Enter your email address in the **User Name** field. Under **Outgoing Mail Server**, enter smtp.office365.com in the **Host Name**. Hit **Next**. Hit **Save** to save the settings. Now go to Mails (Tap Mail icon on your device), you will see new account listed there. Tap the link to see the emails.

Summary

SharePoint Online provides a quick and easy way to setup sites (both private and public) and take advantage of the rich features offered by SharePoint. These sites can be branded and can look completely different than SharePoint sites. SharePoint sites are modern sites that can display elegantly on any device, desktop or mobile. SharePoint Online development lets developers extend out-of-the-box features.

Chapter 12

SharePoint Branding

In this chapter:

- Change SharePoint look and feel
- Use themes
- Add new features using custom development

You have seen how SharePoint works in the online environment in the previous chapter. Now, let's explore how you can brand SharePoint and create custom branded website for your online public website. To begin, you first need to design a site in the program of your choice, for example, Photoshop is popular program to design websites. You can also use Microsoft Expression Web to layout the site in HTML. Design the site and then convert it into HTML. SharePoint 2013 has made it easy to publish custom sites. An HTML site comprises of the html, css and javascript. SharePoint 2013 takes the HTML you create and automatically generates the master page. This is the most powerful tool introduced in SharePoint 2013. You can further customize the master page after it has been generated by SharePoint. SharePoint allows creating responsive designs and you can target multiple devices. SharePoint lets you target a different design for different devices. There is a new feature called **Device Channels**. You can design sites for multiple browsers. In this chapter, we will focus only on the desktop version of the site but targeting other platforms is easy. For this exercise, I have created an HTML site that you can download from the book's website. Once you have downloaded the HTML site, you can start following the steps below to create a master page and a custom look and feel for your site.

1. Login to your SharePoint Online site. Go to the public website and then go to **Site Actions** and select **Site Settings**. Click on **Design Manager**. **Design Manager** lets you change the look and feel of your site and your site does not look like a SharePoint site. Once you have created a custom design, you can use **Design Manager** to convert it into a master page. It is a new feature introduced in SharePoint 2013. It makes it easier to create a fully customized, pixel perfect design. It is available in publishing sites because

developing custom sites is something that is part of the publishing feature. It is available in both SharePoint Server 2013 and Office 365. It is the tools that you can use to manage all aspects of a custom design. The easiest way to upload and work on design files is to map a drive on your computer to the Master page Gallery of your site. This way you can work on the files directly on your local computer.

2. Before we start working on the design, first we have to map a local folder. Add your public website URL to trusted sites in your browser. Go to **Tools > Internet Options > Security > Trusted Sites** and then click **Sites** button. Copy URL to the text box and click **Add**. Click **Close** and then **Ok**.

3. If you are logged into your site, log out and then login again and make sure you check **Keep me signed in** check box.

4. Click on **Design Manager** in **Site Settings**. Click **Upload Design Files**.

Figure 12-1: *Upload Design Files*

Copy the link given on this screen. Make sure you highlight the URL by dragging the pointer over the link from start to end and then copy it, either by selecting **Copy** from the context menu or using Ctrl-C. Do not copy the shortcut from the context menu. You have to map your local folder to this online location. The steps below are for a Windows 7 machine. Steps for other operating systems may be slightly different. Open Computer by clicking the **Start** button and then clicking **Computer**. Click **Map network drive**.

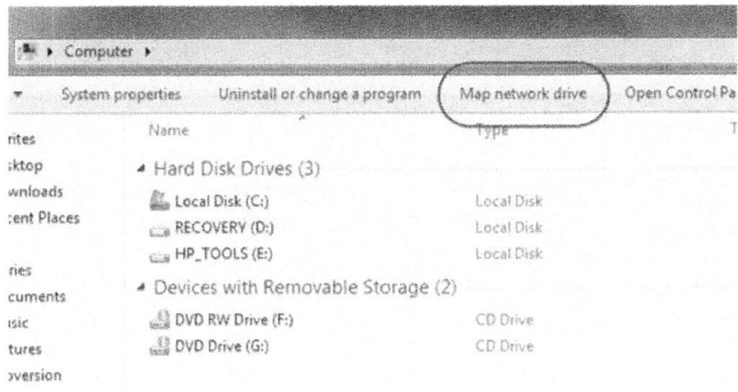

Figure 12-2: *Map network drive*

5. In the **Drive** list, click a drive letter. Choose any available letter.

6. In the Folder box, paste the URL you copied above. You can also type the path manually.

7. Select the **Reconnect at logon** check box. Also, check the **Connect using different credentials**.

8. Click **Finish**. Enter your username (email) and password and check **Remember my credentials** check box and click **Ok**. This will create your mapped folder. If you get an error message, make sure you have copied the correct URL in the trusted sites in the browser. If everything seems ok but you are still denied access, then there is one option to overcome the hurdle but this is not recommended. In Internet Explorer, go to **Tools > Internet Options > Security > Trusted Sites** and then click **Custom level…** button at the bottom. At the bottom, in **Reset custom settings**, from the drop down, select **Low** and click **OK**. Click **Ok** again to close the options. Close all browser windows and open a new browser window and login to your public website and then repeat the mapping process described above and you should be able to map the folder this time.

9. Now copy your HTML site folder to the mapped folder. Your HTML site should have following folder structure:

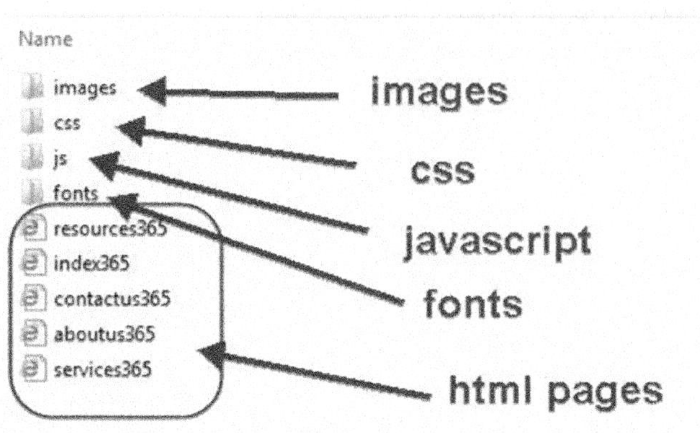

Figure 12-3: *HTML site folder structure*

For this exercise, you can download the HTML site from the download section of this book's website. Unzip the folder to your local hard drive. It will have a site folder called **Social**. The folder structure of this site is shown in the above figure. Copy the whole folder (Social) to the mapped drive. You may have already noticed that the mapped network drive points to the master page gallery of your site. It has other out-of-the-box master pages also. Our new master page will reside in the **Social** folder on the server. We only need the index file to create the master page. Other pages will be created directly on the new site.

You need to be careful about one thing. The index page HTML should be XML compliant and should not have any missing tags or typos otherwise the conversion of HTML to master page may fail or you may get errors and warnings. In the **Status** column, you will see **Warnings and Errors**. Please note, even if apparently there is nothing wrong with the html file, you must validate it before trying to convert it. If you open it directly in the browser, it will work absolutely fine but **Design Manager** checks your page even for the tiniest of issues and reports errors if there are issues in the page, for example, if a starting or ending tag is missing, or a page directive is missing at the top, and so on. If you do get errors during conversion, click on **Warnings and Errors** link to view the issues. Fix the issues reported here and then try to convert the file again. Best way to fix all issues is to use an XML validator and then fix the issues before uploading the HTML file to the server. There are many free XML validators available online. Many sites let you check your pages free of cost online.

10. Now that we have copied our HTML site to the master page gallery on the server, we are ready to convert the index file to the master page. Go to **Site Actions > Site Settings**. Click on **Design Manager** link. Click **Edit Master Page** link on the left.

11. Click **Convert an HTML file to a SharePoint master page** link.

12. Locate **Social** folder and click it. Select index file (index365.html) and click **Insert** button. Click **Publish** button when you are prompted. It won't make any difference if you chose to not publish it yet. Within few seconds, the new html file will appear alongside other html files. A master page with the same name will also be created in the same folder. SharePoint modifies the HTML file and add various snippets. The ASP.NET master page contains all the required SharePoint controls. **Design Manager** is intelligent enough to ignore the ASP.NET code in the HTML page. Once the master page has been created, you should never edit it directly. If you need to make changes, make them in the HTML file. HTML file is linked to the master page. Any changes made in the HTML file automatically reflect in the master page. If you want to use the master page directly, first you have to disassociate the HTML file from the master page. To disassociate the HTML file, go to **Site Actions > Site Settings** and click **Design Manager** and then click **Edit Master Pages**. Click on ellipsis after the index365 file and again click ellipsis to open the menu and select **Edit Properties**.

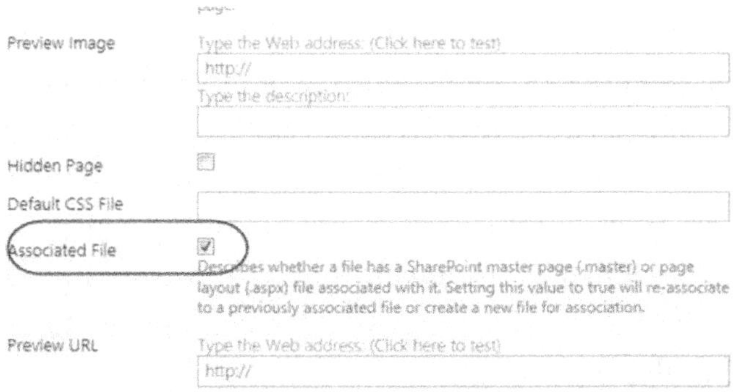

Figure 12-4: *Disassociate HTML file from Master page*

Click **Save** button to save the changes.

13. If you want to see how your site looks with this new design, apply the new master page to the site. First publish the master page. Go to **Site Actions > Site Settings** and click on **Design Manager** link. Click **Edit Master Pages** on the left side. Click ellipsis next to the index page to open the context menu.

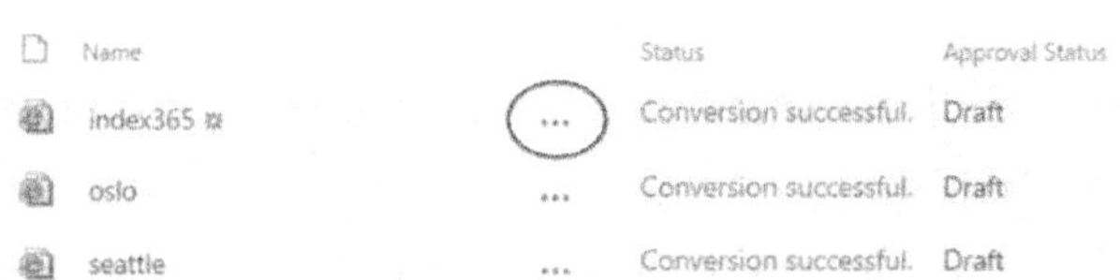

Figure 12-5: *Publish master page*

Again click ellipsis to open the menu and select **Publish a Major Version**. Enter some comments in the **Comments** box, for example, the version number or date and click **Ok**.

14. Go to **Site Actions > Site Settings** and click **Master Page**. In **Site Master Page** section, from the **Default** drop down, select the new master page (social/index365). In **Any Mobile Device** drop down, select the new master page. Do the same in the **System Master Page** section. Make sure second option is selected and from **All Channels** drop down, select new master page. Expand **Alternate CSS URL** section and make sure third option is selected and then click **Browse...** button to locate **main.css** file in the **Social/CSS** folder. Click **Ok** to save the settings. If the css file is not published, you will be prompted to publish it first. Click **OK** to publish it. Once all the changes have been made, It is important to publish all files otherwise they won't show up on the site. For now, you will see the files because you are in Edit mode but it is important to publish all files before making the site live. To publish, go to **Social** folder in **Master Page Gallery** and publish all files in all folders. For now, you can continue editing the site without publishing the files.

15. Preview the home page in browser. You should be able to see the site. There is a JQuery slider under the top menu. The search box is messed up. Menu links are slightly misaligned. This is how the page looks at this time:

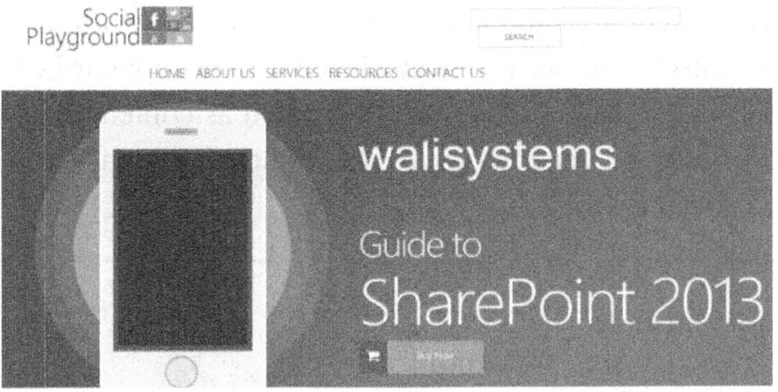

Figure 12-6: *Home page using the new master page*

Now the master page has been created and the home page is up and running. Now in the next section we will edit the master page and add/edit page elements. We will also create the remaining site pages.

Completing the Design

So now you have a master page, next task is to clean it up. As mentioned earlier there are two ways to edit the master page. One is to disassociate it from the HTML page and then edit, the other is to edit the HTML page and that will automatically update the master page. You will choose the second option, that is, to edit the HTML page. You can edit it in SharePoint Designer or any other text editor. For example, for simple tasks I use Notepad++. You can also use Notepad if you want but obviously using SharePoint Designer is a better option since it gives you multiple views to edit the page. You get a code view, and a design view. So open the master page in the editor of your choice. To open it in SharePoint Designer, open SharePoint Designer and browse to the mapped network drive and click **Open**. Then from the **Social** folder, right-click index365.html file and select **Check Out** and then again open the context menu and select **Edit in Advanced Mode**.

You have three sections in the index file: Header, Content and Footer. Header section contains the logo, the search box, the navigation menu and a slider. Content area contains some dummy text and images. Footer contains the navigation menu and two icons, one for facebook and the other for twitter. First thing that you are going to do is remove the content from the content section in the master page and copy it to a text file. You will later add this content to the index page.

1. In the open master page, locate **<div class="fullwidth slider_bg" …** tag. This is where the content area starts. Cut the text that starts with this tag and cover all the text until the footer div starts. Keep it on a clipboard or save in a text file. If you want to save it in a text file, save it as contentarea.txt. Full listing of the code that you need to remove from the master page and copy to the index page is given below:

```
<div class="fullwidth slider_bg"
xmlns="http://www.w3.org/1999/xhtml">
          <div class="main_wrapper">
            <div class="inner_wrapper">
              <div class="slider">
                <ul class="bxslider">
                  <li>
                    <div class="banner_icon">
                      <img src="images/phone.png"
width="100%" alt="" />
                    </div>
                    <div class="banner_content margintop50">
                      <div class="wali_logo">
                        <img src="images/walisystems.png"
width="100%" style="max-width:349px" alt="" />
                      </div>
                      <div class="banner_title white
margintop50">
                        <p class="fsize50 letterspacing2">Guide
to
                        </p>
                        <p class="fsize90 letterspacing3
margint15">SharePoint 2013
                        </p>
                        <a href="#">
                          <div class="button">
                            <div class="icon">
                              <i class="fa fa-shopping-cart fa-
2x">
                              </i>
```

```
                                  </div>
                                  <div class="button_text">Buy Now
                                  </div>
                               </div>
                            </a>
                         </div>
                      </div>
                   </li>
                   <li>
                      <div class="banner_icon">
                         <img src="images/i-pad.png"
width="100%" alt="" />
                      </div>
                      <div class="banner_content margintop50">
                         <div class="wali_logo">
                            <img src="images/walisystems.png"
width="100%" style="max-width:349px" alt="" />
                         </div>
                         <div class="banner_title white
margintop50">
                            <p class="fsize50 letterspacing2">Guide
to
                            </p>
                            <p class="fsize90 letterspacing3
margint15">SharePoint 2013
                            </p>
                            <a href="#">
                               <div class="button">
                                  <div class="icon">
                                     <i class="fa fa-shopping-cart fa-
2x">
                                     </i>
                                  </div>
                                  <div class="button_text">Buy Now
                                  </div>
                               </div>
                            </a>
                         </div>
                      </div>
                   </li>
                   <li>
                      <div class="banner_icon">
```

```
                              <img src="images/mac.png" width="100%"
alt="" />
                          </div>
                          <div class="banner_content margintop50">
                              <div class="wali_logo">
                                  <img src="images/walisystems.png"
width="100%" style="max-width:349px" alt="" />
                              </div>
                              <div class="banner_title white
margintop50">
                                  <p class="fsize50 letterspacing2">Guide
to
                                  </p>
                                  <p class="fsize90 letterspacing3
margint15">SharePoint 2013
                                  </p>
                                  <a href="#">
                                      <div class="button">
                                          <div class="icon">
                                              <i class="fa fa-shopping-cart fa-
2x">
                                              </i>
                                          </div>
                                          <div class="button_text">Buy Now
                                          </div>
                                      </div>
                                  </a>
                              </div>
                          </div>
                      </li>
                  </ul>
              </div>
          </div>
      </div>
  </div>
<div class="main_wrapper">
      <div class="inner_wrapper">
          <div class="content">
              <h1>Lorem Ipsum</h1>
              <p>Lorem Ipsum is simply dummy text of the printing and
typesetting industry. Lorem Ipsum has been the industry's standard
```

```
dummy text ever since the 1500s, when an unknown printer took a
galley of type and scrambled it to make a type specimen book. It has
survived not only five centuries, but also the leap into electronic
typesetting, remaining essentially unchanged.</p>
            <p><img src="images/HTML5.png" width="100%"
style="max-width:550px" alt=""/></p>
        </div>
  </div>
  </div>
<div class="fullwidth grey">
 <div class="main_wrapper">
        <div class="inner_wrapper">
            <div class="sharepoint_book_banner">
             <img src="images/banner.png" width="100%" alt=""/>
             </div>
          </div>
        </div>
 </div>
```

Listing 12-1: *Content for the index page*

2. Locate the div called ContentPlaceHolderMain. See figure below and notice the highlighted text. The outside div is called “ContentPlaceHolderMain”. Inside div is called “DefaultContentBlock”. Delete the inside div **DefaultContentBlock** as it is just a place holder. Cut the highlighted text (after deleting the inner div DefaultContentBlock) and place it just before the footer div (<div class="footer" xmlns="http://www.w3.org/1999/xhtml">).

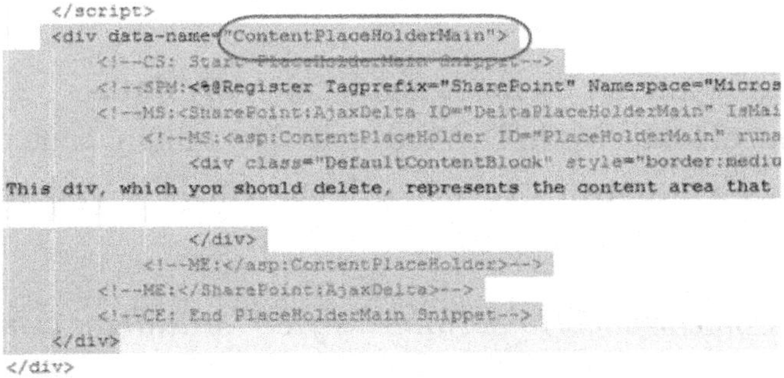

```
    </script>
    <div data-name="ContentPlaceHolderMain">
        <!--CS: Start PlaceHolderMain Snippet-->
        <!--SPM:<%@Register Tagprefix="SharePoint" Namespace="Micros
        <!--MS:<SharePoint:AjaxDelta ID="DeltaPlaceHolderMain" IsMai
            <!--MS:<asp:ContentPlaceHolder ID="PlaceHolderMain" runa
                <div class="DefaultContentBlock" style="border:mediu
This div, which you should delete, represents the content area that

                </div>
            <!--ME:</asp:ContentPlaceHolder>-->
        <!--ME:</SharePoint:AjaxDelta>-->
        <!--CE: End PlaceHolderMain Snippet-->
    </div>
</div>
```

Figure 12-7: *ContentPlaceHolderMain*

ContentPlaceHolder is a .NET control that may be overwritten by the content in the page layout.

3. Save the master page and preview it in browser to test your changes. Now, we will create the first page of the site, index page.

4. Go to **Site Actions > Site Contents** and select **Pages** library.

5. From the **FILES** tab at the top, select **New Document**. Enter Title and Description, for example, for home page, enter **Home** in the **Title** box and enter **Home Page** in **Description** box. Enter **default** in the **URL Name**. Select **Full page** in the **Page Layout** section and click **Create** button.

6. Click on default page you just created to open it. Click **EDIT** link in the top right corner. In the middle of the page, you will see **Page Content** area. This is the place where you will copy the content stored in the contentarea.txt file (or from the clipboard if you did not store it in the text file). Click **Edit Source** button in the ribbon. Copy the content in the **Page Content** area. Preview the page. You will notice that images are missing. This is because the image paths are incorrect in the content you pasted. Click the **Edit Source** button in the ribbon. Notice the image path, it points to the images folder inside Pages library. This is not where you have your images. You can either move your images to the location used in the content or you can change the path in the content to where you have stored your files. Let's use the second option. Replace /**Pages/images** with /**_catalogs/masterpage/social/images**. Save the changes and preview the page. It looks good now, isn't it?

7. Open the master page again in Edit mode and locate **<div data-name="ContentPlaceHolderMain">**. Add style="text-align: center;" in the tag to center-align the content area to match our template. Updated tag will look like this:

   <div data-name="ContentPlaceHolderMain" style="text-align: center;">
   </div>

8. Now, you will create the remaining pages for the site. Refer to the instructions above to create new pages. Create a new page and name it **Aboutus**. Edit it and paste following code in the **Page Content** area (Use **Edit Source** button in the ribbon to switch to code view):

```
<div class="fullwidth slider_bg">
    <div class="main_wrapper">
            <div class="inner_wrapper">
                    <div class="slider">
            <img
src="/_catalogs/masterpage/social/images/aboutus.png"
width="100%"/>
        </div>
      </div>
    </div>
</div>

<div class="main_wrapper">
        <div class="inner_wrapper">
          <div class="content">
            <h1>Lorem Ipsum</h1>
            <p>Lorem Ipsum is simply dummy text of the printing and
typesetting industry. Lorem Ipsum has been the industry's standard
dummy text ever since the 1500s, when an unknown printer took a
galley of type and scrambled it to make a type specimen book. It has
survived not only five centuries, but also the leap into electronic
typesetting, remaining essentially unchanged.</p>


        </div>
   </div>
   </div>

</div>
```

Listing 12-2: *About Us page content*

Save and Publish the page.

Figure 12-8: *About Us*

Create **Services** page (services.aspx) and paste following code in the **Page Content** area:

```
<div class="fullwidth slider_bg">
        <div class="main_wrapper">
                <div class="inner_wrapper">
                        <div class="slider">
                <img
src="/_catalogs/masterpage/social/images/services.png"   width="100%"/>
        </div>
     </div>
   </div>
</div>

<div class="main_wrapper">
        <div class="inner_wrapper">
           <div class="content">
              <h1>Lorem Ipsum</h1>
              <p>Lorem Ipsum is simply dummy text of the printing and
typesetting industry. Lorem Ipsum has been the industry's standard dummy
text ever since the 1500s, when an unknown printer took a galley of type
and scrambled it to make a type specimen book. It has survived not only
five centuries, but also the leap into electronic typesetting, remaining
essentially unchanged.</p>
        </div>
```

```
    </div>
    </div>

</div>
```

Listing 12-3: *Content for Services page*

Create page called **Resources** (resources.aspx) and paste following code in the **Page Content** area:

```
<div class="fullwidth slider_bg">
        <div class="main_wrapper">
                <div class="inner_wrapper">
                        <div class="slider">
                <img
src="/_catalogs/masterpage/social/images/resources.png"  width="100%"/>
        </div>
      </div>
    </div>
</div>

<div class="main_wrapper">
        <div class="inner_wrapper">
          <div class="content">
            <h1>Lorem Ipsum</h1>
            <p>Lorem Ipsum is simply dummy text of the printing and
typesetting industry. Lorem Ipsum has been the industry's standard dummy
text ever since the 1500s, when an unknown printer took a galley of type
and scrambled it to make a type specimen book. It has survived not only
five centuries, but also the leap into electronic typesetting, remaining
essentially unchanged.</p>

        </div>
    </div>
    </div>

</div>
```

Listing 12-4: *Content for Resources page*

Create page called **ContactUs** (contactus.aspx) and paste following code in the **Page Content** area:

```
<div class="fullwidth slider_bg">
        <div class="main_wrapper">
                <div class="inner_wrapper">
                        <div class="slider">
                <img
src="/_catalogs/masterpage/social/images/contactus.png" width="100%"/>
        </div>
      </div>
    </div>
</div>

<div class="main_wrapper">
        <div class="inner_wrapper">
          <div class="content">
             <h1>Lorem Ipsum</h1>
             <p>Lorem Ipsum is simply dummy text of the printing and
typesetting industry. Lorem Ipsum has been the industry's standard dummy
text ever since the 1500s, when an unknown printer took a galley of type
and scrambled it to make a type specimen book. It has survived not only
five centuries, but also the leap into electronic typesetting, remaining
essentially unchanged.</p>


        </div>
    </div>
    </div>

</div>
```

Listing 12-5: *Content for ContactUs page*

That is it! We have created all the pages.

Changing Styles

You can customize the style of the pages by changing the style sheets. Styling is provided by the core files. You can overwrite default styles in your CSS. For example, if home page is loaded in the IE, press F12 to open the debugger. Press **Ctrl+B** to select element by click (This option is available under the **Find** menu). Click the element for which you want to change the style. Debugger shows you the file name where the style is located and the style itself. Copy the style to your custom CSS and apply new setting. For example, below you will change the color of the Suite Bar.

9. Load home page in IE. Press **F12**. Select **Select element by click** from the **Find** menu and select the **Suite Bar**.

Figure 12-9: *Suite Bar*

10. Note the style under the **Style** tab.

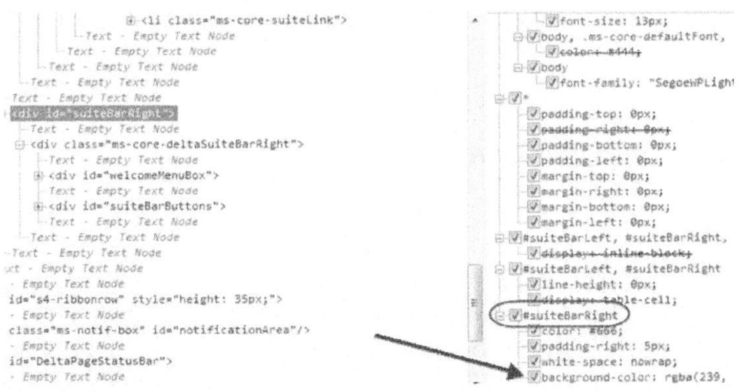

Figure 12-10: *Styles*

To change the color of the Suite Bar, copy following code to the custom CSS file (main.css), at the bottom. You can access css file from the mapped folder.

```
#suiteBarLeft {
background-color: #008ba9;
}
#suiteBarRight {
background-color: #008ba9;
}
```

Listing 12-6: *Custom styles*

Before applying new style, this is how Suite Bar looked:

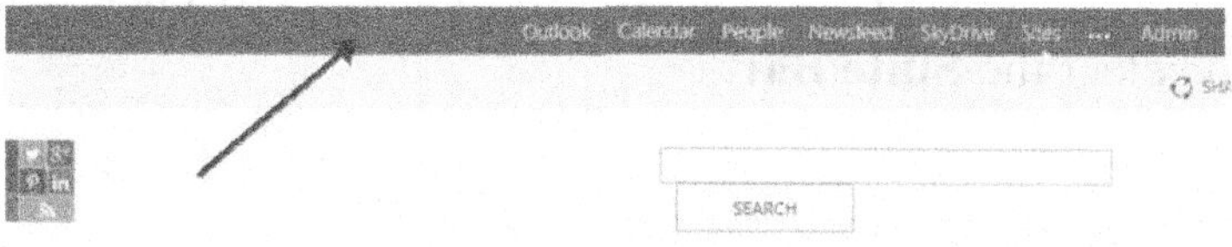

Figure 12-11: *Original Suite Bar Color*

This is how updated Suite Bar looks (You won't notice the color change if you are reading a printed book because the images are mono color):

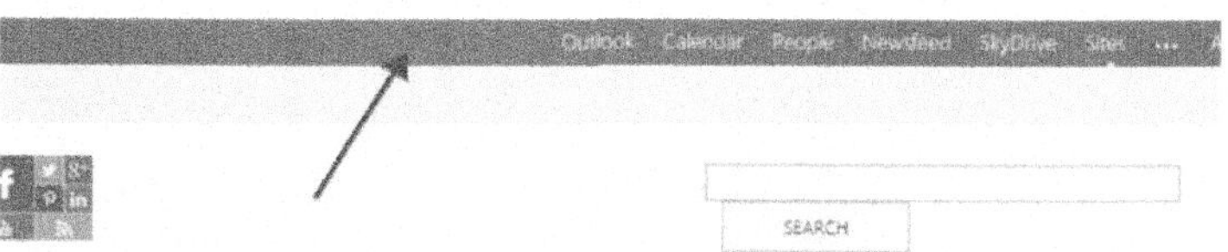

Figure 12-12: *New Suite Bar Color*

Adding Snippets

A snippet is an HTML representation of a SharePoint component or control such as a navigation bar or a Web Part. Using the snippet gallery, you can quickly add SharePoint functionality to your pages. Using snippet gallery, you can do the following:

- Choose a component from the gallery
- Configure the properties of the component
- Preview the appearance in the browser
- Copy it anywhere in your page

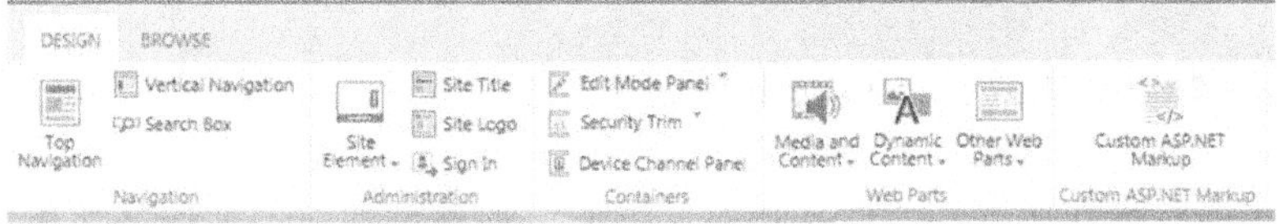

Figure 12-13: *Snippets (Ribbon)*

1. To insert a snippet, browse to the website.
2. Go to Site **Actions > Site settings > Design Manager**.
3. Click **Edit Master Pages** on the left side.
4. Click on the default page to open it.
5. Click on the **Snippets** link in the upper-right corner.
6. Open **Design** tab to view all snippets. Once you click a snippet, the page reloads and shows you a preview of that snippet, the properties for that snippet, and the HTML code snippet that you can copy and paste into your master page. The properties are grouped into different tabs. Expand each tab to see the properties contained within. You can set values for the properties. After you configure properties, choose **Update** to update the HTML snippet. The HTML snippet is automatically updated. Once you are done changing the properties' values, you can copy the HTML and paste it on to your master page. Click **Copy to Clipboard** to copy the snippet to the clipboard.

Customizing Search Box

Now that you have seen how snippets work, let's customize the Search Box in our site. In the HTML template, we had just used a text box and a button. Now, we need a functional search box that should return results from the site. To achieve this, we will add **Search Box** snippet to our master page.

1. To insert **Search Box** snippet, browse to the website.

2. Go to Site **Actions > Site settings > Design Manager**.

3. Click **Edit Master Pages** on the left side.

4. Click on the default (master page) to open it.

5. Click on the **Snippets** link in the upper-right corner.

6. Select **Search Box** snippet from the ribbon.

7. Since you will use custom style, you don't have to change the properties. Click **Copy to Clipboard** button.

8. Go to mapped network drive and open master page in an editor.

9. Locate **<div class="search">** tag.

10. Copy the **Search Box** snippet between **<div class="search">** and **</div>** tags overwriting whatever is in between. Here is how the div tag looks before copying the snippet:

```
<div class="search">
            <input type="text" class="search_box" />
            <input type="button" class="search_button" value="Search"
/>
</div>
```

Listing 12-7: *Search code HTML*

After copying snippet, search div looks like this:

```
<div class="search">
<div data-name="SearchBox">
<!--CS: Start Search Box Snippet--><!--SPM:<%@Register
Tagprefix="SharePoint"
Namespace="Microsoft.SharePoint.WebControls"
Assembly="Microsoft.SharePoint, Version=16.0.0.0, Culture=neutral,
```

```
PublicKeyToken=71e9bce111e9429c"%>--><!--
MS:<SharePoint:DelegateControl runat="server"
ControlId="SmallSearchInputBox">--><!--PS: Start of READ-ONLY
PREVIEW (do not modify)--><div class="ms-webpart-chrome ms-
webpart-chrome-fullWidth "><div WebPartID="00000000-0000-0000-
0000-000000000000" HasPers="true" id="WebPart" width="100%"
class="ms-WPBody noindex " OnlyForMePart="true"
allowDelete="false" style="">
<div componentid="ctl00_SmallSearchInputBox_csr"
id="ctl00_SmallSearchInputBox_csr"><div id="SearchBox"
name="Control"><div class="ms-srch-sb ms-srch-sb-border"
id="ctl00_SmallSearchInputBox_csr_sboxdiv">
<input type="text" value="Search..." maxlength="2048"
accessKey="S" title="Search..."
id="ctl00_SmallSearchInputBox_csr_sbox" autocomplete="off"
autocorrect="off"
onkeypress="EnsureScriptFunc('Search.ClientControls.js', 'Srch.U',
function() {if
(Srch.U.isEnterKey(String.fromCharCode(event.keyCode)))
{$find('ctl00_SmallSearchInputBox_csr').search($get('
ctl00_SmallSearchInputBox_csr_sbox').value);return
Srch.U.cancelEvent(event);}})"
onkeydown="EnsureScriptFunc('Search.ClientControls.js',
 'Srch.U', function() {var ctl =
$find('ctl00_SmallSearchInputBox_csr');ctl.
activateDefaultQuerySuggestionBehavior();})"
onfocus="EnsureScriptFunc('Search.ClientControls.js', 'Srch.U',
 function() {var ctl =
$find('ctl00_SmallSearchInputBox_csr');ctl.hidePrompt();
ctl.setBorder(true);})"
onblur="EnsureScriptFunc('Search.ClientControls.js', 'Srch.U',
function() {var ctl = $find('ctl00_SmallSearchInputBox_csr');
if (ctl){ ctl.showPrompt(); ctl.setBorder(false);}})" class="ms-
textSmall ms-srch-sb-prompt ms-helperText" /><a title="Search"
class=
"ms-srch-sb-searchLink"
id="ctl00_SmallSearchInputBox_csr_SearchLink"
onclick="EnsureScriptFunc('Search.ClientControls.js', 'Srch.U',
function() {$find('ctl00_SmallSearchInputBox_csr').search($get(
'ctl00_SmallSearchInputBox_csr_sbox').value);})" href="javascript:
{}"><img
src="/_catalogs/theme/Themed/E97CFE0A/searchresultui-
```

```
61174269.themedpng?ctag=4" class="ms-srch-sb-searchImg"
id="searchImg" alt="Search" /></a></div></div></div><noscript><div
id="ctl00_SmallSearchInputBox_noscript">It looks like your browser
does not have JavaScript enabled. Please turn on JavaScript and try
again.</div></noscript><div
id="ctl00_SmallSearchInputBox"></div><div class="ms-
clear"></div></div></div><!--PE: End of READ-ONLY PREVIEW--
><!--ME:</SharePoint:DelegateControl>--><!--CE: End Search Box
Snippet-->
</div>
</div>
```

Listing 12-8: *Search snippet*

The code above looks jumbled up. The portions that are of your interest have been emboldened. Since the search box that we had used was just a place holder and did not add any value to the site, we will keep the search box created by the snippet above but that does not mean you cannot modify the look and feel of the search box. You can change the look and feel by customizing the styles. As explained earlier in the chapter, use **F12** key to open IE debugger, select the elements you want changed, copy the styles to your custom CSS and modify them. To clarify the process, I will slightly modify the custom CSS for the search box. Since we are using the snippet generated by the SharePoint, we don't need the text box and search button styling that we had in our CSS, therefore, we are going to remove those styles from our CSS.

11. Open main.css (from the mapped drive > social > css folder) in editor. Locate the following two lines and remove them since we don't need them anymore but even if you keep them they won't harm you since they are not being called any more.

```
.search_box{width:68%; float:left; background-color:#fdfdfd;
border:1px solid #e5f0f3; padding:3% 1%; border-right:0; font-family:
'SegoeWPLight';}

.search_button{width:28%; float:left; background-color:#008ba9;
border:none; color:#FFF;padding:3% 1%; text-transform:uppercase;
font-family: 'SegoeWPLight'; cursor:pointer}
```

Listing 12-9: *Search styling*

The style that we are interested in is as following:

```
.search{width:40%; float:right; margin-top:15px}
```

Listing 12-10: *Search box adjustment*

Since we have removed the Search button and now we are using the default image (lens), either we need to increase the length of the search box or we need to slightly move it to the right to bring it closer to the right border. We will decrease the width from 40% to 25% and that will move the search box closer to the right border. Updated style looks as following:

```
.search{width:25%; float:right; margin-top:15px}
```

Listing 12-11: *Updated* search box *adjustment*

That is it. We have our search box and this is what we will keep on the site. If you want to experiment a bit more for your learning please continue reading. If you want to change the style of the search box, for example, background color, change square corners to rounded corners, etc, copy following style to your custom CSS just under the **.search** class.

```
#SearchBox .search, #SearchBox .ms-srch-sb-border, #SearchBox .ms-srch-sb-borderFocused {
        background-color: #e5f0f3;
        float: right;
        height: auto;
        width: auto;
        padding: 0 0 0 2px;
        -moz-border-radius: 5px;
        -webkit-border-radius: 5px;
        -khtml-border-radius: 5px;
        border-radius: 5px;
        margin-right:10px;
}
```

Listing 12-12: *Stylize search box*

This style will change the background of the search box to light sky blue and square corners to rounded corners. This is how the search box will look:

Figure 12-14: *Stylish Search Box*

If you want to add more style to the search box, for example, if you want to change the search image, add following styles to your custom css (main.css):

```
#SearchBox .ms-srch-sb input[type=text]{
        background-image:url(../images/search_bg.png);
        background-repeat:repeat-x;
        height:20px;
        padding-top:4px;
        width:190px;
        border:none;
        float:left;
        margin:2px 0 0px 0;
        color:#d1d3d4;
        padding-left:10px;
}

#SearchBox .ms-srch-sb-searchImg {
        height:30px;
        margin-left: -19px;
   width: 28px;
        border:none;
        float:left;

}

.ms-srch-sb > .ms-srch-sb-searchLink {
background-image:url(../images/search_btn.png);
        background-repeat:repeat-x;
        height:30px;
margin-left: -19px;
   width: 28px;
```

```
        border:none;
        background-color:transparent;
                float:left;
                cursor:pointer
}
```

Listing 12-13: *More styles for the search box*

In the above styles, we change the background color of the search box to bluish silver and add a gradient to it. We also change the image button to a more stylish button that has a bigger lens image and has stylish border. The images used in the above styles are in the images folder.

Figure 12-15: *More styles for the search box*

I have commented out the above styles in the main.css. If you want to use them, just uncomment the code and feel free to add your own customizations.

Adding Sign-In

Here is another snippet that you can add in your master page. Sign In link is automatically displayed on your page in the upper left corner. You can customize the look and feel and change its position to display it anywhere you like. The Sign In snippet is a security trimmed control. You can add a new Sign In snippet and remove the one that is part of the master page by default or you can customize the one that is already on the page. Below we will look at both options. To add a new snippet, repeat the steps explained previously to add a snippet to the master and choose Sign In snippet. Copy the HTML to clipboard and return to the master page (in editor) and paste the snippet where you want it displayed and then remove the default snippet from the page. For your convenience, I am listing the steps again below:

1. Go to Design Manager and select Edit Master Pages link on the left side.
2. Click the HTML master page (index365.html).
3. Click the Snippets link in the top right corner.

4. Click Design tab.
5. Click Sign In snippet in the ribbon.
6. Click Copy to Clipboard button to copy the snippet HTML.
7. Paste the snippet to the master page at your desired location.
8. Locate following code in the master page and remove it. This is the default Sign In link code.

```
<!--MS:<SharePoint:SPSecurityTrimmedControl runat="server"
AuthenticationRestrictions="AnonymousUsersOnly">-->
   <!--MS:<wssucw:Welcome runat="server"
EnableViewState="false">-->
   <!--ME:</wssucw:Welcome>-->
<!--ME:</SharePoint:SPSecurityTrimmedControl>-->
```

Listing 12-14: *Default Sign In Link*

Now, the second option is to re-use the default Sign In link. All you can do with it is add some style and change its location otherwise it works perfectly fine. Depending on the design of your site, you may want to put it in an appropriate location. To add styling to the Sign In link, you have to add a style class to the CSS and embed the code shown in the above listing within a div with that style class.

9. Locate the code shown in the previous listing and add **<div class="signin_btn">** before it and **</div>** after it. Updated code looks like this:

```
<div class="signin_btn">
<!--MS:<SharePoint:SPSecurityTrimmedControl runat="server"
AuthenticationRestrictions="AnonymousUsersOnly">-->
<!--MS:<wssucw:Welcome runat="server" EnableViewState="false">-->
<!--ME:</wssucw:Welcome>-->
<!--ME:</SharePoint:SPSecurityTrimmedControl>-->
</div>
```

Listing 12-15: *Sign In Link*

10. Now cut the whole snippet (as shown in the listing above) by pressing Ctrl+X (or **Cut** command from the context menu)

11. Locate following line in the master page and paste the code (from the clipboard) immediately after it:

```
<div class="main_wrapper" xmlns="http://www.w3.org/1999/xhtml">
```

Listing 12-16: *Location of Sign In link*

After moving the Sign In snippet to new location, code looks like this:

```
<div class="main_wrapper" xmlns="http://www.w3.org/1999/xhtml">
<div class="signin_btn">
<!--MS:<SharePoint:SPSecurityTrimmedControl runat="server"
AuthenticationRestrictions="AnonymousUsersOnly">-->
<!--MS:<wssucw:Welcome runat="server"
EnableViewState="false">-->
<!--ME:</wssucw:Welcome>-->
<!--ME:</SharePoint:SPSecurityTrimmedControl>-->
</div>
<div class="inner_wrapper">
<div class="logo">
```

Listing 12-17: *Sign In link new location*

By default, Sign In link appears in the top left corner. We have moved it to the right corner just above the Search box and have add some style to it. Class **signin_btn** is in the main.css and looks like this:

```
.signin_btn{
  float:right;
  font-size:12px;
  text-transform:uppercase;
  color:#49176d;
  margin-right:10px !important;
  margin-top:5px !important;
```

```
  padding:0;
  margin:0
}

.signin_btn span a {
   display: inline;
   float: right;
   line-height: 0 !important;
   margin-bottom: -19px;
   margin-left: 0 !important;
   margin-right: 0 !important;
   margin-top: 9px;
   padding: 0 !important;
}
```

Listing 12-18: *Sign In link styling*

Publishing the site

Now, you are reading to publish the site. Publishing the site is not difficult but there are several small steps that must be completed before the site is published.

1. Go to **Site Actions > Site Settings > Design Manager** and click **Publish and Apply Design** link on the left side. In the first paragraph, second line, there is a link **Go to the Master Page Gallery**, click it.

2. Go to **social** folder. You will see four subfolders: css, fonts, images, js and five or six files. Go to each folder one by one and publish all the files within each folder. On the root folder (social), publish the master page (index365.html). I will show you how you can publish the master page. Follow the same procedure to publish all other files in all folders. Hover mouse over index365.html and click the down-arrow that appears to open

the menu, and select **Publish a Major Version**. Add some comments in the **Comments** box and click **OK**.

Final Steps

Final steps include setting up the navigation links (in header and footer), adding facebook and twitter links in the footer, setting up SEO, page titles, and so on.

1. From the mapped drive, open master page in an editor.

2. Change page names in the navigation (menu2 – see below).

```
<nav id="menu2" class="menu2">
    <ul>
        <li>
            <a href="index.html">Home

            </a>
        </li>
        <li>
            <a href="about-us.html">About Us

            </a>
        </li>
        <li>
            <a href="services.html">Services

            </a>
        </li>
        <li>
            <a href="resources.html">Resources

            </a>
        </li>
        <li>
            <a href="contactus.html">Contact Us
```

Figure 12-16: *Page names*

3. Replace index.html with ../../../Pages/default.aspx, about-us.html with ../../../Pages/aboutus.aspx, services.html with ../../../Pages/services.aspx, resources.html with ../../../Pages/resources.aspx, contactus.html with ../../../Pages/contactus.aspx. Note that new pages are located in the **Pages** library and master page is located in the **_catalogs/masterpage/social** folder, therefore, you have to give correct path to access the pages.

4. Locate **<div class="menu_footer">** and replace page names with new ones under this tag. Replace index.html with ../../../Pages/default.aspx, about-us.html with ../../../Pages/aboutus.aspx, services.html with

../../../Pages/services.aspx, resources.html with ../../../Pages/resources.aspx, contactus.html with ../../../Pages/contactus.aspx.

5. In the same footer tag, there are Twitter and LinkedIn icons, add URL to your accounts in the **href** (Replace # sign with the URL).

```
<li>
    <a href="#">
        <img src="images/Twitter.png" alt="" />
    </a>
</li>
<li>
    <a href="#">
        <img src="images/LinkedIn.png" alt="" />
    </a>
</li>
```

Figure 12-17: *Social media links*

6. Locate **<div class="social_icons">** tag and add URL to your Twitter and LinkedIn account in the href (replace # with the URL).

```
<div class="social_icons">
    <a href="http://twitter.com/sharepointjobs2">
        <img src="images/Twitter.png" alt="" />
    </a>
    <a href="http://www.linkedin.com/company/hot-sharepoint-jobs">
        <img src="images/LinkedIn.png" alt="" />
    </a>
</div>
```

Figure 12-18: *Social Icons*

7. Save your changes and re-publish the master page. All your links should be working now.

8. Open home page in browser. Click **SITE** tab and click **Edit Title** button in the ribbon. If you have not already set Title and Description, add them now in the **Title** and **Description** boxes and click **Ok**.

9. Click **Change Logo** button and provide path to the logo. You can upload logo from the computer or select it from the SharePoint site if you already have it up there.

10. Clic **Edit Properties** and select **Edit SEO Properties**.

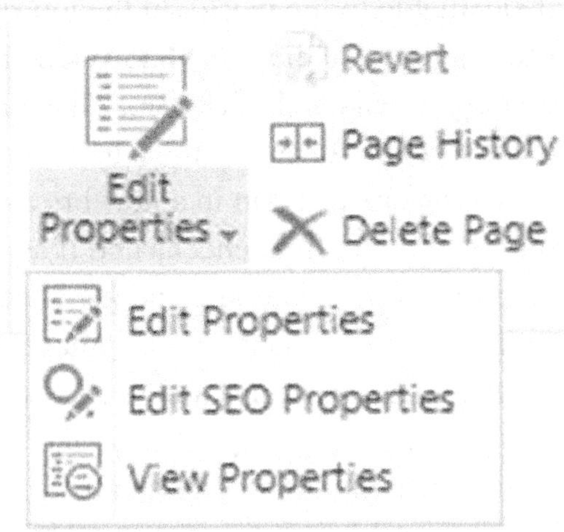

Figure 12-19: *SEO Properties*

11. **Title** field should already be populated. Title comes from the page's title tag. If you want to change it, you can change it on this page. This it the title that search engines use in their search results. To change Browser Title, enter a title in the **Browser Title** field. This is the title that appears in the title bar of the browser. **Meta Description** field contains the description of your page. Search engines display this description under your link in the search results. Make sure the description is concise and meaningful and contain some of your keywords. In the **Keywords** field, add keywords that are relevant to your home page. Keywords are comma-separated and you can add a maximum of 1024 characters in this field. Finally, you can choose to include this page in the sitemap and expose it to the search engines. If you do not wish to include this page in the sitemap, choose **Exclude from sitemap and hide from Search Engines** otherwise keep the first option selected. Click **Ok** to save the changes. Repeast the process for all other pages.

12. Go to **Site Actions** > **Site Settings** > **Welcome** Page. Click **Browse** button and locate your default page in the Pages library and click **OK**. With this step, you have set your site's homepage. Users now don't have to type complete URL in the browser, they just enter the domain name and home page automatically loads.

13. Final step is to make the website available to the world. In the top right corner, under the Suite Bar, there is a button called **MAKE WEBSITE ONLINE**. Click this button, enter comments and click **OK**. That's it. Your website is now live. Wait for a few seconds for the settings to take effect. Logout and close browser. Open your website again in a new browser and you will see your site up and running. Do the final checks and if something has been left, login and fix it. Start promoting your site and generate traffic. SharePoint public site is a goldmine for small businesses as you get all useful features like document management, meetings and events scheduling, email management, contacts management, tasks management and so on with your public website that acts as an informational brochure and introduces your business to the world. It is an affordable solution for small businesses that do not want to spend extravagantly on the development of such a useful website that can actually play a role in the promotion and management of their business. SharePoint Online is also available for medium and large sized businesses. If you are an enterprise, buy an enterprise package. It is still cheaper than hosting a site inhouse because the infrastructure and hardware costs are quite high and then you also need resources to handle and maintain your site(s). With a hosted solution, you can handle the day to day tasks yourself or at most you need one administrator to keep the site in good health.

Summary

SharePoint 2013 has made it easy to publish custom sites. An HTML site comprises of the html, css and javascript. SharePoint 2013 takes the HTML you create and automatically generates the master page. This is the most powerful tool introduced in SharePoint 2013. You can further customize the master page after it has been generated by SharePoint. SharePoint allows creating responsive designs and you can target multiple devices. SharePoint lets you target a different design for different devices.

Chapter 13

SharePoint Online (Office 365) Development

In this chapter:

- Setup development environment
- Create Hello World application
- Publish apps

SharePoint Online provides a robust, flexible development environment for developers to create custom solutions. SharePoint Online is a cloud based service, hosted by Microsoft, for businesses of all sizes. Instead of installing and deploying SharePoint server on premises, businesses can simply subscribe to SharePoint Online to provide their employees with an enterprise grade solution to create sites to share documents with colleagues, partners, and customers. SharePoint Online provides the following features:

- Teams can manage documents and files online.
- Teams can collaborate and work on projects online.
- Company sites are a source of information for the employees and sites are available to all employees. Public sites are available to clients and partners.
- SharePoint Online is secure.

SharePoint Online is customizable. It can be customized in three ways:

- You can use browser to do configurations and settings.
- You can customize sites using SharePoint Designer.
- You can develop solutions that extend out-of-the-box features.

Now that you have seen how SharePoint Online looks like and you have learned how to create a public branded website, it's time to learn a few basic things about the development. Development approach to SharePoint Online is scoped to the site collection. There are two development models in SharePoint Online development. One is Sandboxed solutions. It provides a framework for developers to create, upload and activate solutions to SharePoint Online. These solutions run in an isolated environment that has access to a core subset of the server object model. Developers have access to the major objects at site collection level. The other model is Client object model. This model provides three representations of the core objects in the server side model: (i) .NET framework managed model (ii) Silverlight model (iii) ECMAScript model. Client object models provide remote access to SharePoint data and functionality.

In this chapter, you will learn how to setup the development environment and develop your first "Hello World" application.

If you already have access to the Online environment then you are ready to explore the environment for development. If you haven't tried it before, you can sign-up for a trial account that gives you access for 30 days. If you have MSDN subscription, you get full 1-year access to SharePoint Online and you can add development tools for free. These development tools are called as "Napa" Office 365 development tools. "Napa" tools are a great way to get started building apps for Office or SharePoint directly out of a browser window. There is no need to install any other tools such as Visual Studio. All you need is an Office 365 account and a browser. To get started, sign up for an Office 365 developer site. Then, install "Napa" tools. To sign up go to the following page and click **Sign up for an Office 365 Developer Site**.

http://msdn.microsoft.com/en-us/library/office/fp179924.aspx

There are three ways to get access to the developer site. You need to have an MSDN subscription. Second option is to purchase an enterprise subscription. Third option is to get a free 30-days trial. Sign-up process is easy and takes just a few minutes. Once you have signed up, provisioning takes some time to complete and after the servers have been provisioned, you can start developing apps.

Figure 13-1: *Activate subscription in MSDN*

Once you have signed up. Go to your site and log in. Your site URL will look like this:

http://walisystems.sharepoint.com

Remember, to login you have to use your email ID and password, for example, your user ID will be username@sitename.onmicrosoft.com.

Developing Hello World Application

1. Once logged in, open SharePoint admin center. From menu, open **Admin** drop down and select SharePoint.

Figure 13-2: *SharePoint admin center*

2. Before you can start developing, you must create a new site collection or you can use the one created by default. To create a new site collection, click **New** button under **Site Collections**. You have to choose from one of the following: (i) Private Site Collection (ii) Public Website. You can choose either of them. Since you are developing a Hello World application, choose **Private Site Collection**.

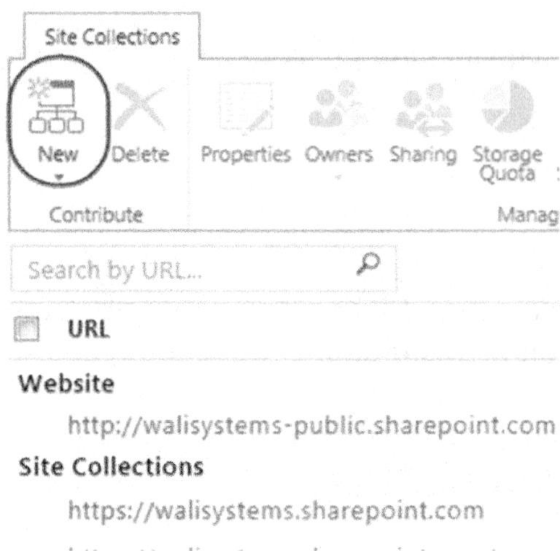

Figure 13-3: *New site collection*

3. Fill out the new site collection form. Click **OK** after completing the form.

Title	OfficeOnline
Web Site Address	Keep the default option selected
Web Site Address (2nd Line)	/sites/**officeonline**
Template Selection	Collaboration > Developer Site
Time Zone	Choose your time zone
Administrator	Enter your name or email and click **Check It** button/icon
Storage Quota	Min. is 100MB so enter 100
Server Resource Quota	300

4. Go to your newly created site collection. Either enter the site URL directly in the browser, for example, https://sitename.sharepoint.com/sites/yoursitename OR click the site collection URL under site collections, site collection properties will open.

Click the link in Web Site Address.

5. Click **Build an app** (1st box) to start the developing an application. Since you still don't have development tools, you will be prompted to add development tools. Click **ADD IT** button. You will have to sign in with your Live ID, the one you used to subscribe to the **Office 365 Developer Site**. You are already signed in with your Office 365 login, so log out first and then login with your Live ID. After signing in, the development tools will be added to your site collection.

Figure 13-4: *Add "Napa" Tools*

6. After the "Napa" tools have been added, you will see **"Napa" Office 365 Development Tools** in the left navigation under **Recent**. You can also access it via **Site Contents**. Click the link to go to the app. You will be redirected to https://www.napacloudapp.com.

7. If this is your first application, you will be asked to choose the kind of app you want to build. If you have developed before, you will see the existing apps and a **Add NewProject** button. If you are not prompted to choose App type, click **Add New Project** button.

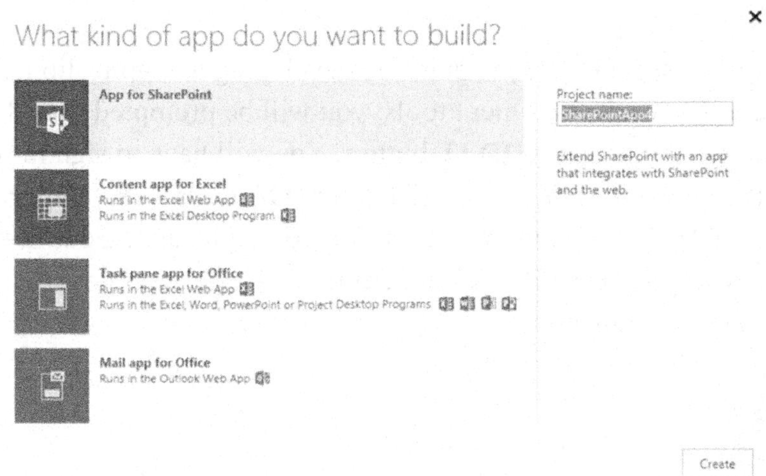

Figure 13-5: *Choose App for SharePoint*

8. Choose **App for SharePoint** and enter application name in **Project name** box. Click **Create.**

9. Default.aspx will open automatically in edit mode. This file contains all the file references that you need to get started. Code editor provides features such as syntax highlighting and code completion. Shortcut keys are also available. To view a list of shortcut keys, press CTRL and the "." (dot) key. You can also open the project in Visual Studio. To see commands, click ellipsis in the right corner and click **Show editor commands**.

Figure 13-6: *Show editor commands*

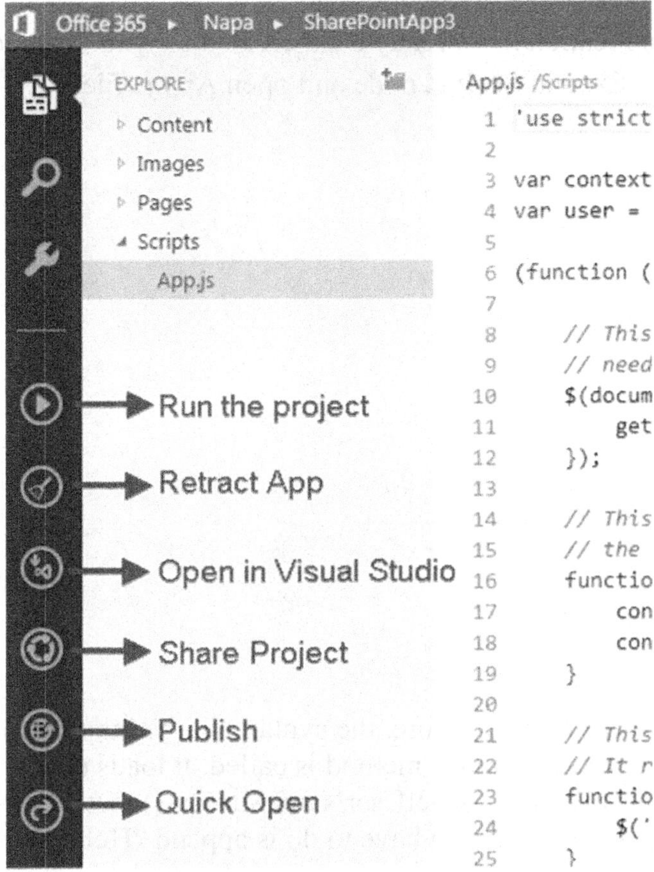

Figure 13-7: *Development environment*

There are 31 commands shown in the list.

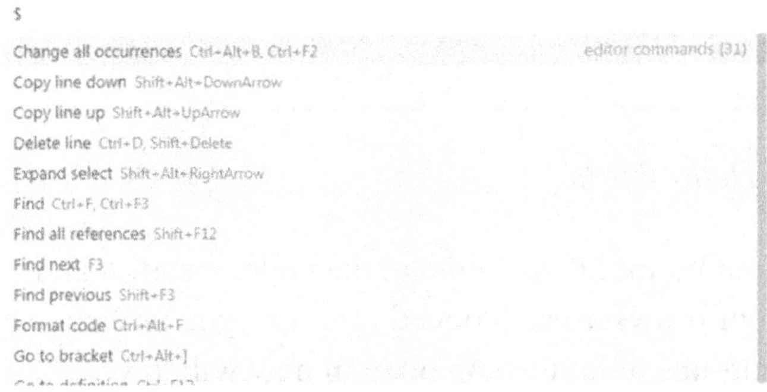

Figure 13-8: *Editor Commands*

10. The Hello World application has already been created for you. You don't have to do anything. Expand **Scripts** node and open App.js file.

```
6  (function () {
7
8      // This code runs when the DOM is ready and creates
9      // needed to use the SharePoint object model
10     $(document).ready(function () {
11         getUserName();
12     });
13
14     // This function prepares, loads, and then executes
15     // the current users information
16     function getUserName() {
17         context.load(user);
18         context.executeQueryAsync(onGetUserNameSuccess,
19     }
20
21     // This function is executed if the above call is s
22     // It replaces the contents of the 'message' elemen
23     function onGetUserNameSuccess() {
24         $('#message').text('Hello ' + user.get_title()
25     }
```

Figure 13-9: *Hello World*

If you have worked with JQuery before, the syntax will be familiar to you. When the app loads, getUserName() method is called. It loads user properties in the context object. onGetUserNameSuccess() method displays the user name on the screen. All you have to do is append "Hello World" after the user name.

```
function onGetUserNameSuccess() {
                $('#message').text('Hello ' + user.get_title() + 'and Hello
World!');
        }
```

Listing 13-1: *Display Hello World*

11. Click Run Application button. It will upload the code, install it and the launch the app. If your browser has blocked pop-ups, you will have to temporarily allow pop-ups to let the app open in new window.

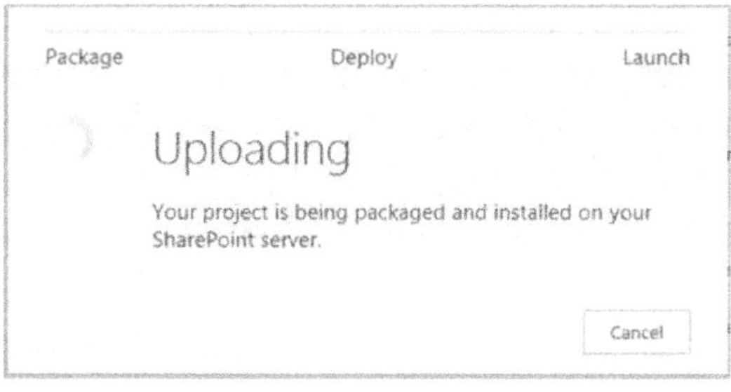

Figure 13-10: *Launching app*

12. The app will launch in a new window.

Figure 13-11: *Running app displaying Hello World!*

13. Click Publish button.

Publish settings

Specify a name and version for the app package. The package will be published to the App Packages library.

Package name

SharePointApp2

Create a public link that can be shared with others

Version

Major Minor Build Revision

1 . 0 . 0 . 0

"Napa" configures apps to support the en-US locale. To support other languages and cultures open this project in Visual Studio.

Publish

Figure 13-12: *Publish settings*

Change Package name to your desired app name. You can create a public link for your app that you can share with others. Click **Publish**.

14. You will see **Publish successful** message. Click **publish location** link.

15. Next thing you would like to do is to publish the app to **App Catalog** so that you can add it to other areas of your site. To add an app, you select **Add an app** from **Site Actions**. Since you have not published it to App Catalog yet, you cannot add it in other site areas. Go to Admin center (From menu choose Admin > SharePoint).

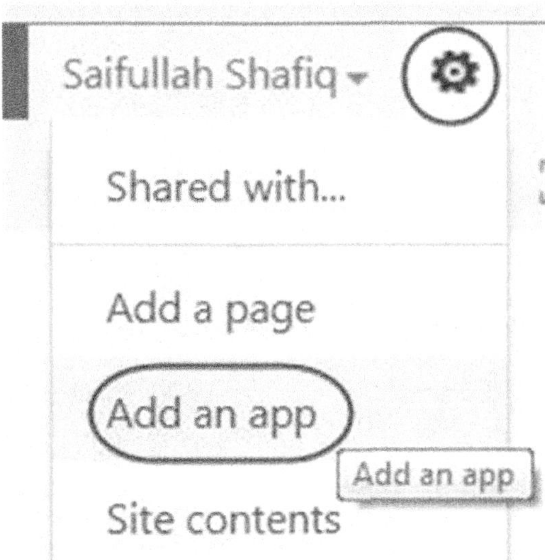

Figure 13-13: *Add an app*

16. Choose **apps** from the menu.

Figure 13-14: *Choose apps*

17. Click App Catalog link.

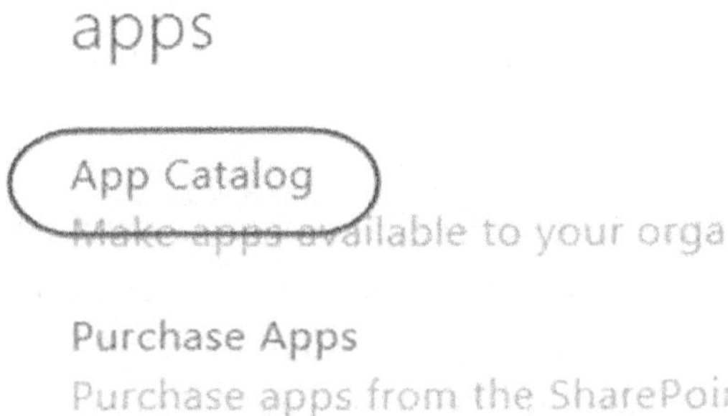

Figure 13-15: *App Catalog*

18. Choose a new catalog if one already does not exist. Choose **Create a new app catalog site** and click **OK**. Creating new app catalog is just a new site collection creation.

19. After app catalog has been created, go back to publish location. You will see you app listed under App Packages. Click your app to download it. Save it on your machine.

20. Now go to app catalog. Click **Distribute apps for SharePoint.**

Figure 13-16: *Distribute apps for SharePoint*

21. Click **new app** link. Click Browse to locate the app you downloaded, select it and click **OK** to upload it to the catalog. Fill out the form. Since this is just a test app, you can leave default values selected and click the Save button. For real-world apps, you will have to provide as much information as possible to list the app in the correct category, you will want to add the Publisher name and so on.

22. Go to a SharePoint site other than the one you developed the app on. The app is already available on the site collection where it was developed. To test adding it from the app catalog, you have to try that on another website. You can choose the main website for this test. Once on the main site, click **Site Actions** and select **Add an app**.

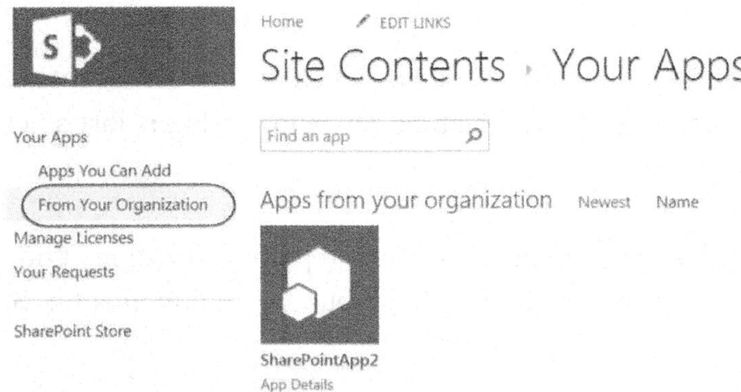

Figure 13-17: *From Your Organization*

23. Click **From Your Organization** in the menu and then click **App Details** link under the app. Click **ADD IT** button to add the app to the site.

24. You will be asked if you trust the app. Click **Trust it** button.

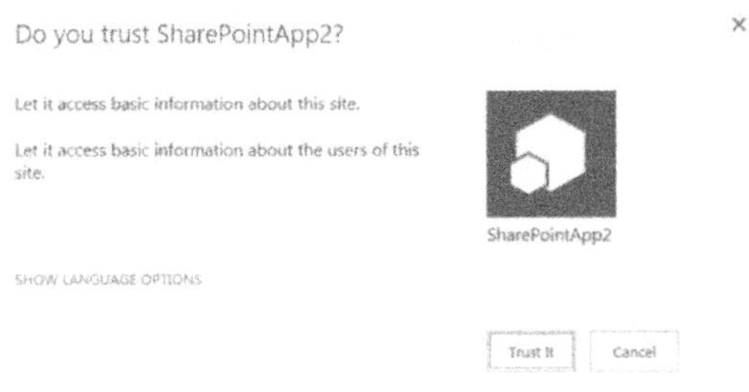

Figure 13-18: *Do you trust the app?*

25. You will see your app listed on **Site Contents** page. Click the app to run it. You will see "Hello *Yourname* and Hello World!" message.

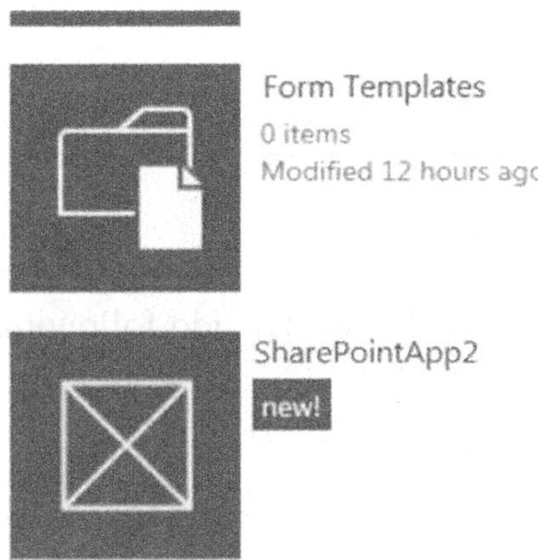

Figure 13-19: *App listed on Site Contents page*

26. You have seen the development and publishing process. We will slightly change the application to add a bit of functionality. We will add a method that will display the number of lists in the site and the names of the lists. Open default.aspx in editor and remove the code from the PlaceHolderMain and add following code:

```
<asp:Content ContentPlaceHolderID="PlaceHolderMain" runat="server">
  <div>
      <button id="getLists">Get Lists</button>
  </div>
</asp:Content>
```

Listing 13-2: *Add button to the page*

This button will call the method that will return the number of lists and their names.

27. Open App.js in editor. Add two new variables at the top:

```
var context = SP.ClientContext.get_current();
var user = context.get_web().get_currentUser();
var web = context.get_web();
var lists = web.get_lists();
```

Listing 13-3: *Define variables*

28. In the main function(), after **getUserNames();**, add following code:

```
$("#getLists").click(function (event) {
        showLists();
        event.preventDefault();
});
```

Listing 13-4: *Call ShowLists()*

29. Add following functions after the main function():

```
function showLists()
{
        lists = web.get_lists();
        context.load(lists);
```

```
        context.executeQueryAsync(onshowListsSuccess,onshowListsFail);
}

function onshowListsSuccess(sender, arg)
{
        var lstEnum = lists.getEnumerator();
        var strLists ='';

        while(lstEnum.moveNext())
        {
                if (strLists==='')
                strLists = lstEnum.get_current().get_title();
                else
                strLists += ', ' + lstEnum.get_current().get_title();
        }
        alert('Total # of lists is: ' + lists.get_count() + ' and List names are: '
+ strLists);
}

function onshowListsFail(sender, args)
{
        alert('Error occurred! Message: ' + args.get_message());

}
```

Listing 13-5: *Add functions*

executeQueryAsync() executes the current pending request asynchronously on the server by using the client side object model (CSOM). It accepts two parameters. **succeededCallback** is a function or a delegate of the method to call if the request executes successfully. **failedCallback** is a function or a delegate of the method to call if the request fails to execute. When a function executes asynchronously, script continues to run without waiting for the server to respond.

lists.get_count() returns the count of lists in the site.
lstEnum.get_current().get_title() where lstEnum is the list enumerator returns the title of each list. We store list titles in a string and then display the string when the list title reading loop has ended.

30. Click **Run Project** (play icon) button to run the app. App will run and display **Get Lists** button. Click it to see the result. An alert message will be displayed with the number of lists and their names.

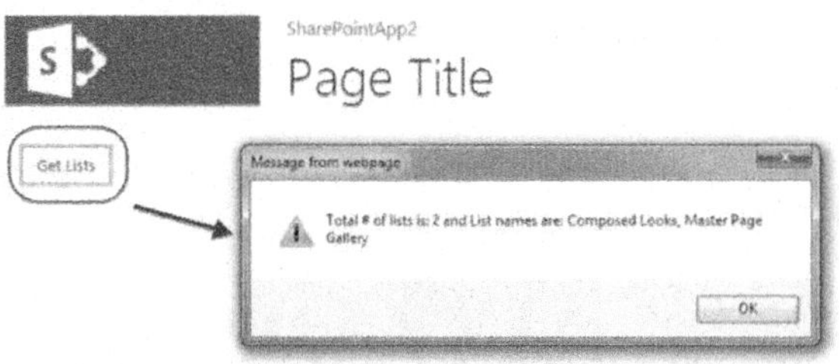

Figure 13-20: *Alert message showing list count and titles*

This brings us to the end of the chapter. You have now acquired the knowledge to develop apps for the cloud. Modern apps use javascript heavily and it may be a good time to brush up your javascript programming skills. The focus of applications is moving from web to mobile and client side scripting provides the necessary boost that modern apps need, it keeps the pages light weight and thus apps can run easily in web as well as mobile environments.

Summary

SharePoint Online provides a robust, flexible development environment for developers to create custom solutions. "Napa" tools can be added to SharePoint Online for developing applications. You get a complete editor in the cloud for development. You can also develop using Visual Studio 2013 with easy publishing features.

Chapter 14

Business Connectivity Services

In this chapter:

- Setting up Business Connectivity Services
- Modifying External Content Types
- Business Data List Web Part

Business Connectivity Services is a centralized infrastructure in SharePoint 2013 and Office 2013 that supports integrated data solutions. SharePoint 2013 acts as an interface that you can use to surface data from external data sources. Enterprises do not store their data in one location. Enterprises use myriad of data sources and applications with each application connected to a separate data source. Before Business Connectivity Services, surfacing data from different sources was not easy and required lot of effort and custom programming. SharePoint 2013 provides an easy way to connect SharePoint interface with Line of Business applications and this make analysis of data very easy and it especially helps the decision makers in the company to perform analysis on the data and thus they can take informed decisions in a timely manner. SharePoint provides data connectors out-of-the-box to connect to external data sources. SharePoint also allows developers to write new connectors to connect to the data sources that cannot be reached with out-of-the-box connectors. Developers can also connect to external sources through web services. Data can be published as an OData source or many other types of external data. The connectors are the communication bridge between SharePoint and the external system that hosts the external data. Reaching external data is one part. Second part is how you display data in SharePoint. SharePoint has several ways to present the external data. Most common way is to use an external list. External lists look and feel like regular SharePoint lists, except that they can only display external data. You can also use out-of-the-box custom list to display external data. In this case, you would external data column. It is a special column that you can add to any list and it is used to display data form an external data source. External data column is a special content type. SharePoint 2013 also includes Business Data Web Parts for presenting and interacting with external data and apps for SharePoint.

Once you have connected to the external data, you can perform different operations on the data. You can create, Read, Update, Delete, and Query on the data. This is a cool feature because you can not only read from the external source but you can also write back using Business Connectivity Services. You can also integrate results from the external source into SharePoint search.

Business Connectivity Services provide an easy way to surface data in one location. The biggest challenge faced by the companies today is dealing with data stored at multiple locations and to access this data companies have to use different systems and mechanisms. It is a big challenge and requires lot of investment and resources. Business Connectivity Services, or BCS for short, has solved this problem. One way is to pull data from external resources and store it inside SharePoint and the other is to keep data where it is but use Business Connectivity Services or Search to look at the data. Most companies have their data stored in File Shares in the form of documents. SharePoint Search allows users to index documents stored in the File Shares. Next chapter discusses search in detail and shows you how you can setup Search for searching file shares. You can also use Business Connectivity Services to search for information stored in external data sources. Companies take data from external sources, and some times these external sources are controlled by third party companies, and make it available to the employees through custom interfaces. This data consists of both structured and unstructured data. With BCS, companies have a way to integrate external data into SharePoint.

Some uses of BCS include applications that are crucial to companies' success, for example, customer support systems, tracking systems (HR, etc), KPIs (dashboards), etc. For example, help desk is a popular application and almost all companies use this application in form or the other. Help desk data may be stored in separate databases. Queries may be stored in one database, knowledge based articles and tips may be stored in another database. To provide a unified system to the help desk technicians so that they can easily look at the queries and can find solutions in the knowledge base database, companies can use BCS. With BCS, companies can retrieve data, filter and sort it and then present it for interaction in an external list in SharePoint. Another popular usage is HR system. Resumes are stored in different formats. Some companies store resumes in the form of files whereas some companies prefer to store candidate data in the database and they use metadata system to identify the candidates easily. Storing data in the database lets companies automate several processes like setting up search agents and automatically connecting with the right candidates when needed. With BCS, connecting to these external databases is easy. A complete HR management system can be developed in SharePoint using Business Connectivity Services.

Business Connectivity Services have an advantage over other solutions and the biggest advantage is its ease of implementation. Custom solutions can be developed using .NET or Javascript but implementing a BCS solution is very easy and does not require any programming unless your requirements are overly complex and advanced.

Business Connectivity Services is a centralized infrastructure that supports integrated data solutions. Infrastructure itself is unaware of any external system. External data is wrapped up in a definition called an external content type. External content types are centrally stored and secured and can be shared by many Business Connectivity Services solutions.

Business Connectivity Services has server-side components and client-side components. Both components use the same configuration data. For the server-side, the configuration data is stored in an external content type. External content types are stored in the Business Data Connectivity (BDC) Metadata Store database. For the client-side, the configuration data is stored in a BDC model on the client in the BDC client-side cache. The BDC model is just a version of the external content type that is exported to an XML file. The XML file is imported into the Office client. There is a lot more that can be discussed in this chapter, for example, how BCS works exactly but that is beyond the scope of this discussion. Below, you will setup an external content type, assign permissions and create an external list to surface data from the external database. You will also see learn to use Business Data List web part.

1. First, you need to setup a database that will act as an external data source. Create a database in SQL Server and name it "Users". Add a table with the following fields and name it "Profiles". Add some records in the table.

Column Name	Data Type	Allow Nulls
ID	int	☐
FirstName	varchar(50)	☑
LastName	varchar(50)	☑
Address	varchar(100)	☑
City	varchar(50)	☑
State	varchar(50)	☑
Zip	varchar(50)	☑
Country	varchar(50)	☑
Phone	varchar(50)	☑

Figure 14-1: *Users table*

2. Open SharePoint site in SharePoint Designer. Click External Content Types under Site Objects.

3. Click External Content Type button in the ribbon to create new content type.

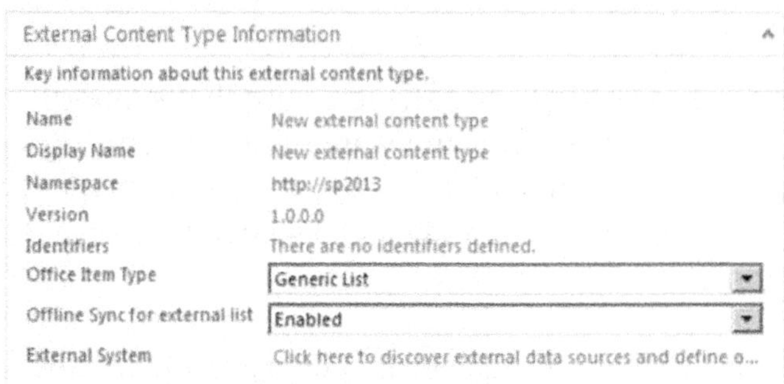

Figure 13-2: *Create new external content type*

4. Type name of the new content type in the Name field. Call it **NewECT.** Click in the **Display Name** field and type **NewECT**.

5. In the field **External System**, click the link **Click here to discover external data sources ….**

6. Click **Add Connection** button.

7. Select **SQL Server** in the **Data Source Type** drop down that appears and click **OK**.

8. You will be prompted to enter database details.

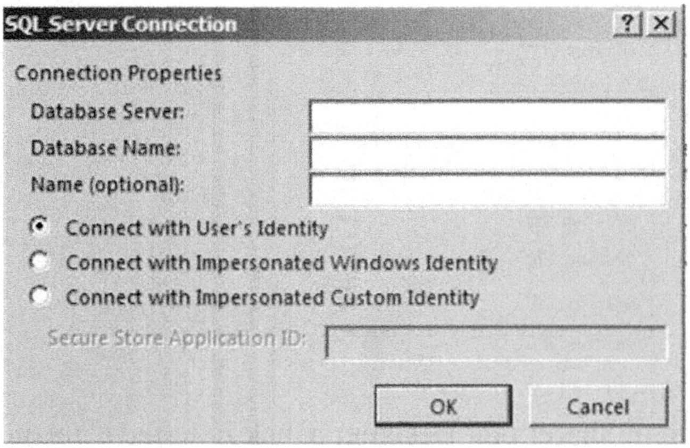

Figure 14-3: *SQL Server Connection*

In the **Database Server** field, enter the name of your database server. In the **Database Name**, enter **Users** (that is the database you created above). Keep the first option (**Connect with User's Identity**) selected. Click **OK.**

9. In **Data Source Explorer** tab, expand **Users** node, expand **Tables** node. Right-click **Profiles** and select **Create All Operations**.

10. Click **Next**.

11. At least one identifier should be specified. In the **Errors and Warnings** box, you will see an error that asks you to select at least one field from the table as an identifier. This identifier uniquely identifies an item in the external system to ensure proper runtime behavior. By defait field **ID** is selected and its properties are shown in the **Properties** box. Check **Map to Identifier** check box. Uncheck **Required** and **Read-Only** check boxes since **ID** field in our database is not unique. It should be a unique field and should be autogenerated. If you opt to keep **ID** field unique, then leave these check boxes checked.

12. We want to show one field in the exteran item picker. Select FirstName from the Data Source Elements and in the Properties, check Show In Picker check box.

13. Click **Next**.

14. Click **Add Filter Parameters** button. In the **Properties** pane, select **Country** in the **Data Source Element** drop down. Click **Click to Add** link in the **Filter** field.

15. Keep all default options and click **Ok**. The **Filter Type** is **Comparison**. The **Operator** is **Equals** and the **Filter Field** is **Country**. In the external list, the filter will default to Country field. User will select a country in the filter column to see records for that country.

16. In **Properties** pane, in **Default Value**, type **US**.

17. Click **Finish**.

18. Click **Save** button in the menu.

19. Click **Create List & Forms** button in the ribbon to create an external list based on the external content type you just created.

20. Type **NewECT** in the **List Name**. If you want to create an InfoPath form for your list, check **Create InfoPath Form** checkbox. I left it unchecked because I wanted the default form to be used. Click OK to save the settings.

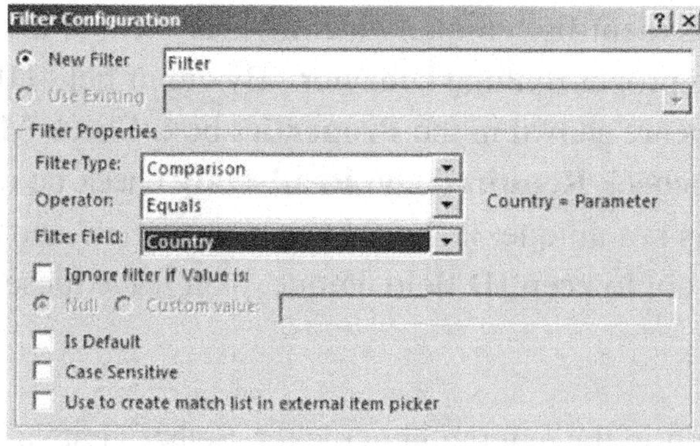

Figure 14-4: *Create new external list*

21. Select **Lists and Libraries** from the left side under **Site Objects**. In the right pane, under **External Lists**, you will see newly created list. Click it to open.

22. Click **Preview in Browser** button in the ribbon to open the list in the browser.

23. You will see **Access denied by Business Data Connectivity** error. There are a few things that you need to be aware of before the list can be previewed in the browser. You have to set permissions in the Central Admin site. Steps below will take you through the whole process for setting up the pre-requisite permissions. Common errors that you get when viewing external list in the browser are as following:

    - BCS Access Denied.

- BCS login failed for user NT Authority

BCS Access Denied

In SharePoint 2013, If you have setup Business Connectivity Services, or BCS for short, and now you are trying to access an external content type in a list but you get "Access denied by Business Data Connectivity", here is you will solve it.

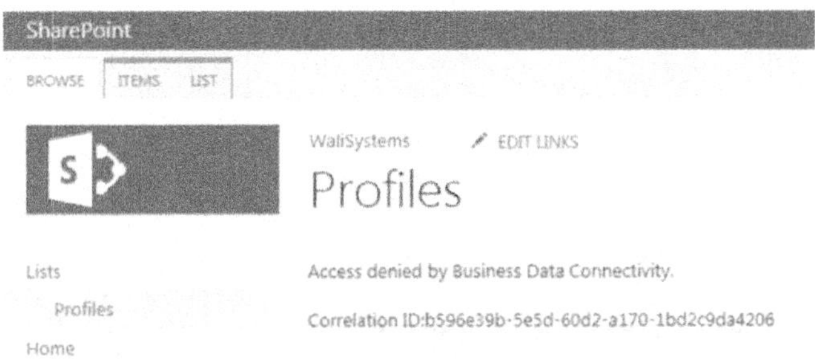

Figure 14-5: *Access denied by Business Data Connectivity*

This error occurs because the logged in user does not have permissions on the external content type that you had created. To resolve the issue go to central administration site and give permissions to specific users on the external content type.

1. Open **SharePoint Central Admin** site.
2. Click **Application Management**.
3. In **Service Applications** section, click **Manage service applications**.
4. Click **Business Data Connectivity Service**.
5. You will see your external content type listed on this page. Select it (by checking the box) and click **Set Object Permissions** button in the ribbon.

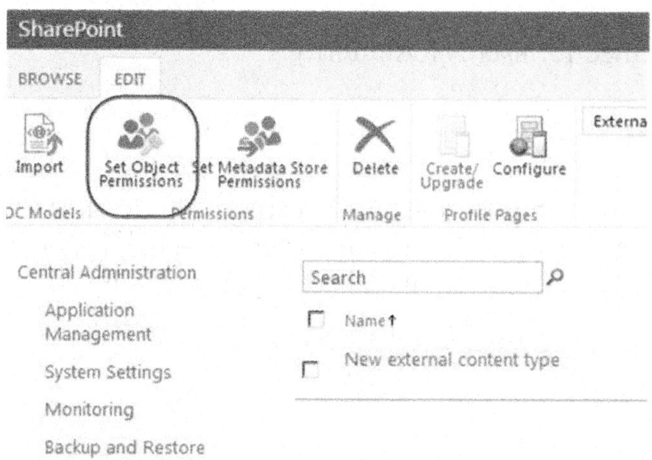

Figure 14-6: *Set Object Permissions*

6. This will open a dialog box. Enter username that you want to give permission to in the box and click **Add**. Select the added user in the box below and select permissions for that user by checking the boxes against the listed permissions. Depending on the type of user, you can select which permissions to give. For example, giving them **Execute** permission will let them see the results. If the user is admin type, then give them all the permissions. Lastly, check the box at the bottom to "propagate" permissions to all methods of the external content type. Click **OK** to save the settings.

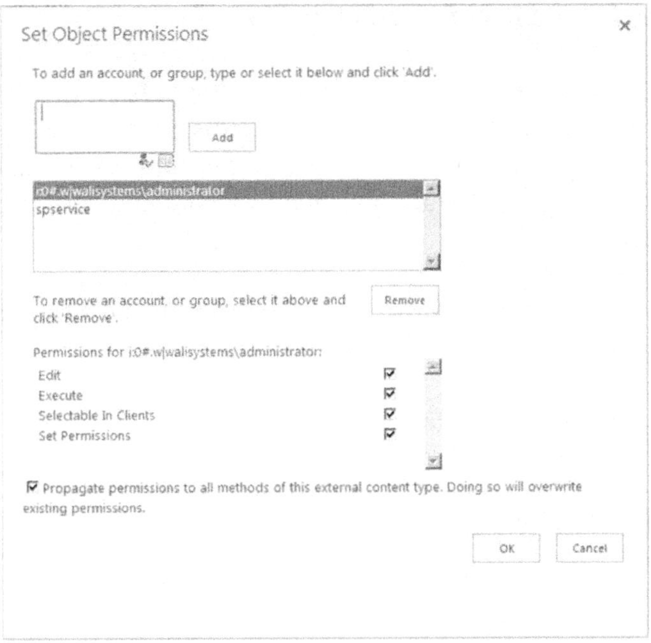

Figure 14-7: *Add an account*

Also, if you want to give access to all users in the organization, add **NT Authority\Authenticated Users** in the box at the top and click **Add** and then assign desired permissions as described above. When you add authenticated users, you see **All Users (windows)** listed as one of the users.

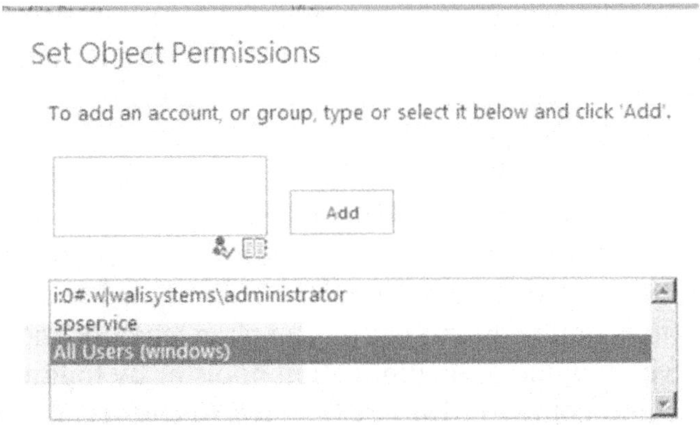

Figure 14-8: *Add All Users (windows)*

BCS Login failed for user NT Authority

If you have created an external content type and are trying to access it in a list and get **Access Denied**, that is because you have not given permissions to the logged in user to use that external content type. You can easily give permissions to that user or to all users in your site by going to central administration site and configuring business connectivity service.

Some users get another very interesting error message after they have assigned permissions on the external content type. The error message is:

Message from External System: 'Login failed for user 'NT AUTHORITY\IUSR'.'.

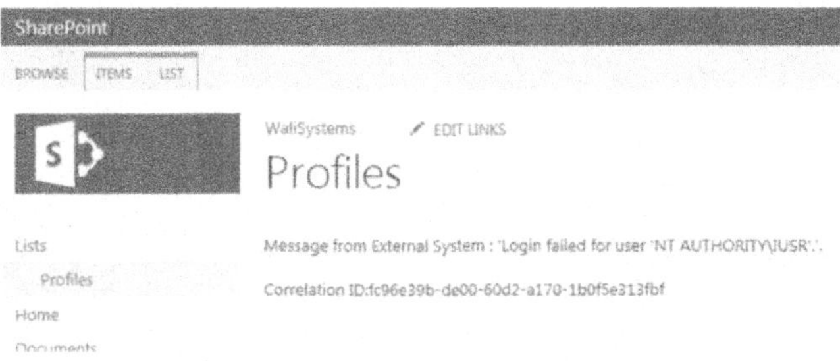

Figure 14-9: *Login failed for user 'NT Authority\IUSR'*

This error occurs because database does recognize the credentials passed from SharePoint. Which database? Database that you are connecting to in your external content type! It depends on how you have setup application pool identity in SharePoint. If you are not using managed service account for your services, you are most likely to get this error. This is not likely to occur in corporate environments but in home or test environments where users use **LOCAL SYSTEM** OR **NETWORK SERVICE** in application pools, this error will occur. This is because **NETWORK SERVICE** translates to **NT AUTHORITY\IUSR** when credentials are passed from web server to database server. **IUSR** is used when user credentials are not available, for example, for anonymous users. To resolve this problem, either change application pool identity or add **NT AUTHORITY\IUSR** to the database permissions. The first scenario is quite common. You can change application pool identity by go to the web server and updating application pools or by logging into SharePoint central admin site and going to **Manage Service Accounts**. We will cover second scenario here, that is, to add **IUSR** to the database.

1. Go to **Microsoft SQL Server Management Studio** and connect to your database server.

2. Expand **Security** node and right-click **Logins** node and select **New Login**.

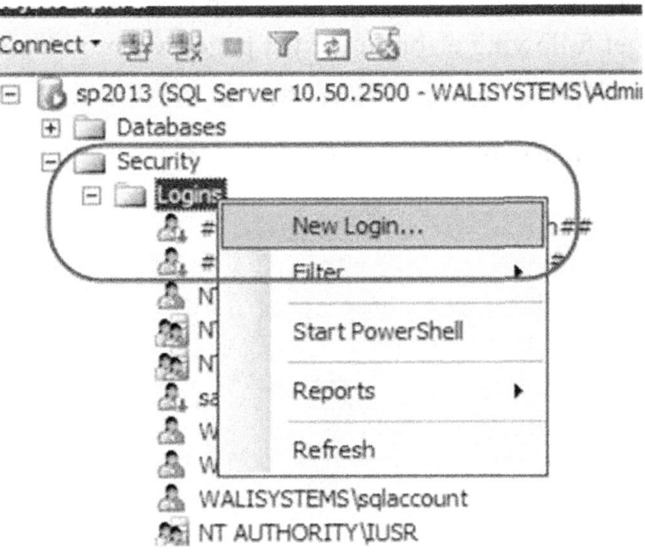

Figure 14-10: *Add New Login*

3. In the **Login Name**, enter **IUSR** and click **Search…**. This will open a new search box. Enter **IUSR** in the object name and click **Check Names**. Click **OK**.

4. You will notice that **Login Name** has been populated with **MACHINENAME\IUSR** where **MACHINENAME** is your machine name, for example, in the figure below, you see **SP2013\IUSR**. **SP2013** is my machine name.

Figure 14-11: *Add IUSR*

This is not what you want to add to the logins. Change **MACHINENAME** to **NT AUTHORITY** so **Login Name** should read **NT AUTHORITY\IUSR**. Click **OK**.

5. Now you may think that the user has been added and your external content type will work. Right? Wrong! You still have to map user to the database otherwise you will get following error on the list page (where you are trying to load external content type):

The query against the database caused an error.

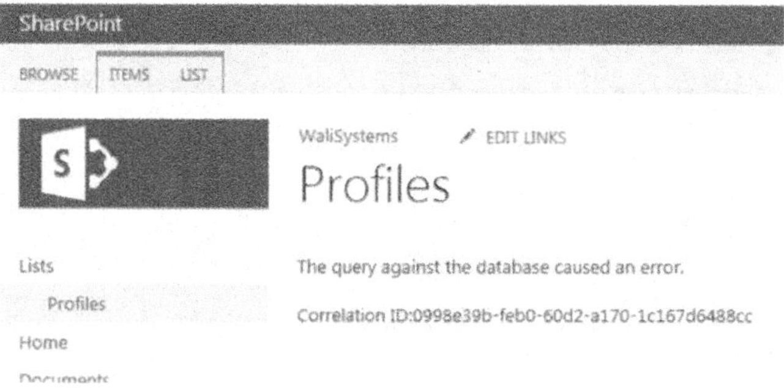

Figure 14-12: *The query against the database caused an error*

This error occurs, as I said, because user is not mapped to the database yet.

6. To map user to the database, right-click **NT AUTHORITY\IUSR** in Logins and select **Properties**.

Figure 14-13: *Login properties*

7. There are two ways to do this mapping. One way is to add user to one of the server roles, for example, **serveradmin** or **sysadmin**. This will give user full rights to all databases and thus you won't have to do explicit mapping. To do this, on the properties page, click **Server Roles** and check **sysadmin**. Click **OK**. Obviously you would not want to do this in real environment. So the other option is to map the user directly to the database that has been used in the external content type.

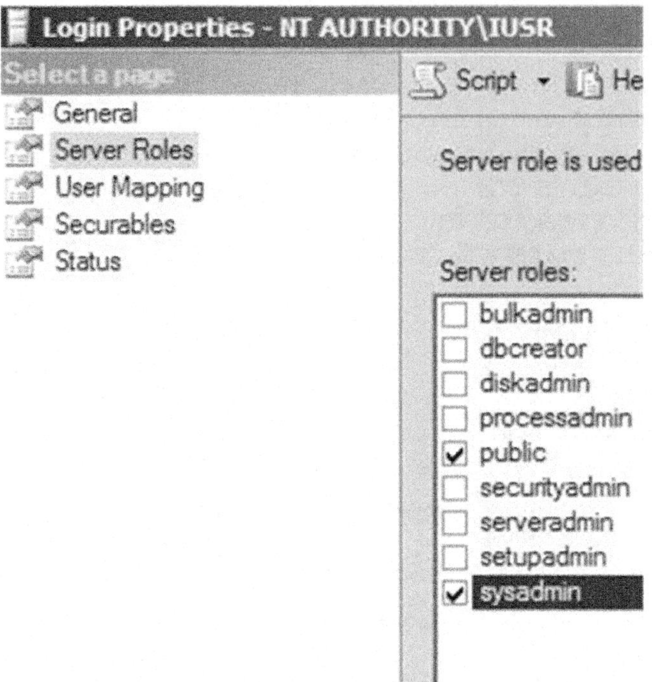

Figure 14-14: *Login properties*

To add user mapping, in **Login Properties**, click **User Mapping**. Locate the database in the list and then check the box in the **Map** column. As soon as you check the box, **NT AUTHORITY\IUSER** appears in the **User** column. That's it. Click **OK** to save the setting. By the way on the same properties page, you can also assign database role membership to the user for the database but that is not required.

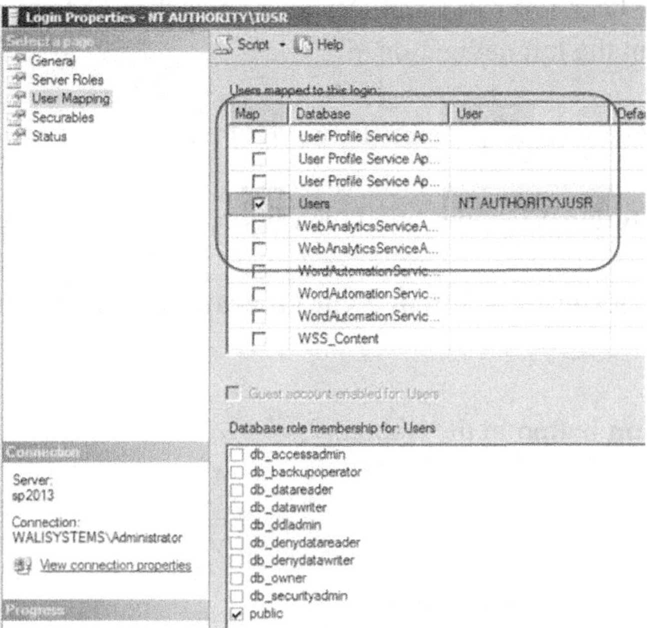

Figure 14-15: *User mapping*

Now, go back to the list page and reload it. You will see results from the external data source.

Figure 14-16: *External list in browser*

Modifying External Content Type

You can modify external content type if you want to make changes. For example, in the above figure you may have noticed that you only see three records where the country is US. What if you want to show all records? You can remove the filter from the content type.

1. To remove filter, go back to SharePoint Designer and select **External Content Types** from the left navigation.
2. Click on the external content type name in the right pane.
3. Click **Operations Design View** button in the ribbon.
4. On the right side, in **External Content Type Operations**, select **Read List** operation.
5. Click **Edit Operation** button in the ribbon.
6. Click **Next** to go to the **Filter Parameters**.
7. Click **Remove Filter Parameter** button.
8. Click **Finish**.
9. Click **Save** button in the menu bar to save the changes. Your changes won't take effect if you don't save them.
10. After updating the external content type, you have to add permissions again in the Central Administration as explained in the previous section.
11. Now view list again in the browser and you will see all records.

NewECT

ID	FirstName	LastName	Address	City	State	Zip	Country	Phone
1	Saifullah	Shafiq	100 St	Frederick	MD	21703	US	3432432432
2	James	Timothy	90 Baldwin Rd	Baltimore	MD	21888	US	234343423
3	Melissa	Theron	200 Rooks St	Washington	DC	20004	US	3242342344
4	Timothy	Dalton	111 Baker St	London		11200	UK	234234444
5	Lynda	Johnson	85 Golden Valley Rd	Melbourne		34324	Australia	324243443

Figure 14-17: *List shows all records*

12. To apply filter on the records, move mouse over Country column name and click the arrow to open the menu.

Figure 14-18: *Apply filter*

Click on the **US**, the list will refresh and show only records for the US.

Business Data List Web Part

Now that you have learned how to setup Business Connectivity Services, you will learn how to use Business Data list web part to show external data on any SharePoint page you want.

1. Open SharePoint page where you want to show external data.
2. Edit page and go to **INSERT** tab.
3. Click the **Web Part** button.
4. Select **Business Data** in **Categories**. Select **Business Data List** in **Parts**.
5. Click **Add** button.
6. Click on **Open the tool pane** link or go directly to the web part properties.
7. Click on **Select External Content Types** button.

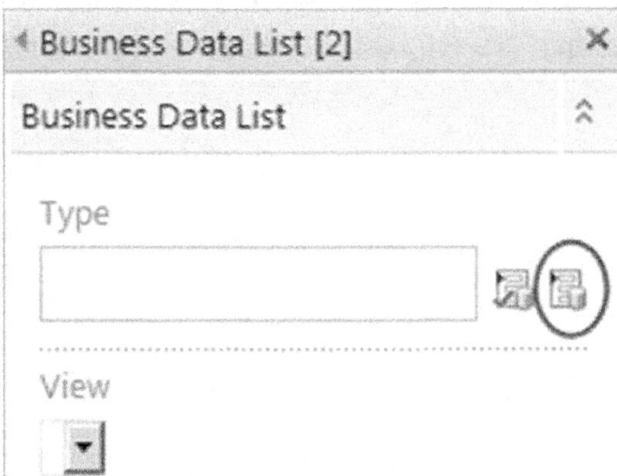

Figure 14-19: *Select External Content Types*

8. You may see the following error in the external content types picker:

The Business Data Connectivity metadata store is unavailable. Check configuration and try again.

This is because you have not setup **Metadata Store Permissions** yet.

9. Go to Central Administration site and under **Application Management** click **Manage service applications**.

Figure 14-20: *Manage service applications*

10. Select **Business Data Connectivity Service** from the list. Do not click on the link but just select the row by clicking anywere on the row except the link. This will highlight the row. Then click **Manage** button in the ribbon.

11. Select the content type that you want to use by checking the checkbox before it and click **Set Metadata Store Permissions** button in the ribbon.

12. Enter the user account that is setting up the web part in the account box and click **Add** button.

13. Select all permissions below by checking all the check boxes. Also, check the checkbox that says **Propagate permissions to all BDC Models …** and click **OK**.

14. Now go back to the SharePoint page, edit the web part properties and select an external content type using the picker. Click **Apply** and then **OK** to close the properties.

15. Click **Save** button to save the page. The web part now shows data from the external data source. This is how the web part looks:

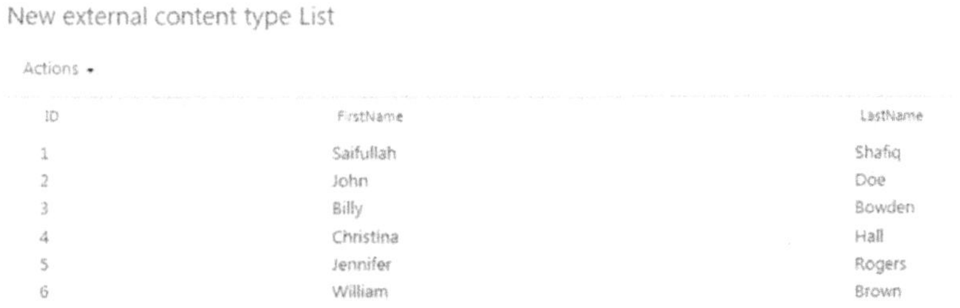

New external content type List

Actions ▾

ID	FirstName	LastName
1	Saifullah	Shafiq
2	John	Doe
3	Billy	Bowden
4	Christina	Hall
5	Jennifer	Rogers
6	William	Brown

Figure 14-21: *External data in a web part*

Summary

Business Connectivity Services is a centralized infrastructure in SharePoint 2013 and Office 2013 that supports integrated data solutions. SharePoint 2013 acts as an interface that you can use to surface data from external data sources. Enterprises do not store their data in one location. Enterprises use myriad of data sources and applications with each application connected to a separate data source.

Chapter 15

SharePoint Search

In this chaper:

- Result Sources
- Federated Search
- Setup Bing Search
- Search External Data Sources
- Setup File Shares Search

SharePoint enterprise search is regarded as one of the world's best search technologies. SharePoint 2013 has gone one step farther and added new enhancements to this world-class technology. There are new enhancements for finding people and expertise. Phonetic search works in multiple languages. Search results display authored documents and information about past projects that can showcase a person's expertise. There are contact cards that provide a way to contact people directly from the search results.

Navigation has been enhanced so that users can find search box quickly. Search results are logically grouped and provide query suggestions. Search center sites can be created that focus on a particular scope. Scopes can be added in the central administration to limit the search to particular areas. By the way, the terminology has slightly changed in SharePoint 2013. Search scopes in SharePoint 2010 are now called as **Result Sources** in SharePoint 2013. Administrator can define a result source. Local SharePoint is the default predefined result source. With a result source, administrator can restrict queries to a subset of content by using a query transformation. You can use a query transformation to return on video results from the local SharePoint index.

One big improvement is the search result preview. Users can rest the pointer over a search result to preview the document content in the **hover panel** to the right of the result.

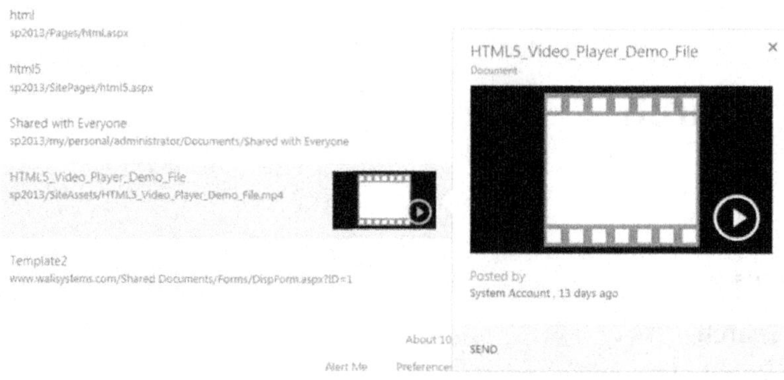

Figure 15-1: *Hover panel*

Certain types of related results are displayed in groups called **result blocks**. A result block contains a small subset of results that are related in a particular way.

Search helps users quickly return to important sites by remembering what they have previously searched.

Site owners can create **result types**. A result type is a rule that identifies a type of result and a way to display it.

Well, these were some of the new enhancements in SharePoint 2013. In this chapter, you will learn how to customize the default SharePoint search interface by adding new search and result pages.

- You will add a new content source.
- You will add a new scope (result source) and add rules for this new scope.
- You will add a new tab in the search center on your SharePoint site and create search pages for this tab. You will create two pages, one that will host the **Search Box** web part and the other will be used to display filtered search results.
- You will modify **Search Box** web part and associate it with the newly created scope. This way you will get filtered search results.

First you will create new result source. This result source will be used in the custom results page that you will create.

Result Source For Custom Search

1. Open SharePoint **Central Administration**.
2. Click **General Application Settings**.
3. Click **Farm Search Administration**.
4. Click **Search Service Application**.
5. On the left hand side, click **Content Sources** under **Crawling**.

Figure 15-2: *Content Sources*

By default, **Local SharePoint sites** content source is already created. If you have not indexed your content before, you can do it now. Hover your mouse over **Local SharePoint sites**, a drop down will appear. Left-click it to open the menu and select **Start Full Crawl** from the menu. If there is no content in your sites, it will take 2-3 minutes for the engine to finish the crawl. You can create new content sources in this section. Once the indexing is complete, you can go and test search in your SharePoint sites. To view crawl log, hover mouse over **Local SharePoint sites** and open menu and then select **View Crawl Log**. Here you can see each content source and number of successful crawls, errors, warnings, top level errors and deletes against each content source.

6. Click **Result Sources** in the left side menu.

Queries and Results
Authoritative Pages
Result Sources
Query Rules
Query Client Types
Search Schema
Query Suggestions

Figure 15-3: *Result Sources (Scopes)*

7. Click **New Result Source**. Enter **New Search** in **Name** field. Enter **New Search Scope** in **Description** field. In **Protocol**, keep **Local SharePoint** selected. In **Type**, keep **SharePoint Search Results** selected. In **Query Transform**, keep default option selected. To build a new query transformation, click **Launch Query Builder**. Remove text from the **Query text** box. Select **Query from the search box** from **Keyword filter** drop down and click **Add keyword filter** button. Select **--Show all managed properties--** from the **Property filter**. Again open **Property filter** and select **Created**. Select **Less than** in the comparison drop down. Select **Today's date (~5 days)** from the value drop down. Click **Add property filter** button. This query will return all items that were created before today's date. Click **OK**.

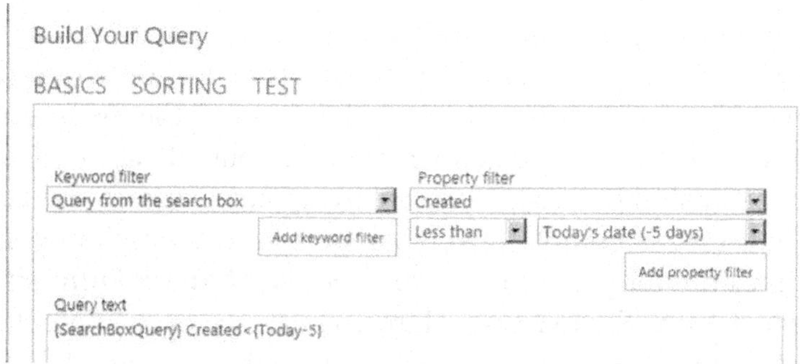

Figure 15-4: *Build Your Query*

In **Credentials information**, keep **Default Authentication** selected. Click **Save** button to save the settings.

8. Go to SharePoint site and create a search center site. Go to **Site Actions >** **Site Contents**. In **Subsites** section, click **new subsite**. In **Title** enter **Custom Search Center**. In **URL name** enter **customsearch**. In **Template Selection** section, select **Enterprise** tab and select **Basic Search Center**. Click **Create** button.

9. Go to the newly created site. Go to **Site Settings > Site Contents**. Go to **Pages** library. If you don't see **Pages** library that's because you have not enabled **SharePoint Server Publishing** feature in **Site Settings**. To enable that feature in site settings, first you have to enable **SharePoint Server Publishing infrastructure** feature in site collection features. Select **FILES** tab. Click **New Document** button. Enter **Custom Search** in **Title**. Enter **csearch** in **URL Name**. Default selection in **Page Layout** is **(Article Page) Body only**. Click **Create**.

10. Add search web part to the page. Click **EDIT** from the top right corner. Click **INSERT** tab. Click **Web Part** button. Select **Search** from **Categories**. Select **Search Box** from the **Parts**. Click **Add** button.

11. Go to web part properties. Select **Send queries to a custom results page URL**. In **Results page URL**, enter **csearchresults.aspx**. If you are not sure about the path, enter complete path, for example, http://site/searchcentersitename/Pages/csearchresults.aspx. Remove the querystring that is appended automatically after the page name. Click **OK**. This page has not been created yet. You will create this page now.

12. Create another page following the instructions given above. Name it **Custom Search Results** and enter **csearchresults.aspx** in the **URL**.

13. Go to the new page. Edit it and add a web part following the instructions given above. This time add **Search Results** web part. After adding the web part, edit its properties. Click **Change Query** button. In the **Select a query** drop down, select the **result source** (New Search (Service)) you created in the previous steps. In the **Keyword filter**, select **Query from the search box**. Click **OK**.

14. Go to the custom search page (csearch.aspx) and try to search for an item. Results will be shown on the custom results page with the results matching the criterion you set in the result source.

WaliSystems
sp2013

Administrator
sp2013/my/personal/administrator

How To Use This Library
dev.walisystems.com/SitePages/How To Use This Library.aspx

AB922B82-8406-4E49-B17B-9057BDF09503
sp2013/Lists/PublishedFeed/AB922B82-8406-4E49-B17B-9057BDF09503

Figure 15-5: *Custom results page*

So, what is the purpose of this exercise? By defining a new results source you have basically filtered the search results. Now, instead of getting all search results, you will get only what is relevant. Remember, when you defined the results source, you added a filter to show you only the items that were created before today's date. By using that particular results source in the **Search Results** web part, you see only filtered results. This is particularly useful in big organizations where each department has a separate site and each site has a huge volume of documents. You can create small search centers that use specific results sources to filter the results.

Federated Search

SharePoint search provides two approaches for processing queries to return search results. One is content crawling and the other is federated search. Let me remind you, if you are an experienced SharePoint user and have used SharePoint 2010, that just like “Search Scopes”, “Federated Search” terminology is extinct in SharePoint 2013. You can still do what you used to do with Federated Search in SharePoint 2010, but that term is non-existant in SharePoint 2013 but regardless concept stays the same. In content crawling, results are returned from the search service application's content index based on the query. The content index contains content that is crawled by the search service application and includes text content and metadata for each content item. In federated search, you are enabled to display search results from sources other than SharePoint. This content is not crawled by your search server. With federation, the query can be performed over the local content index, or it can be forwarded to an external content repository where it is processed by that repository's search engine.

Advantages of using federated search are many. You require no additional capacity requirements for the content index as content is crawled by external entity. You can take advantage of a repository's existing search engine. You can federate to an Internet search engine to search the web. You can access repositories that are secured against crawls but which can be accessed by search queries.

Below, you will learn to set up federated search that will return results from Bing search engine.

Set up Bing Search

1. Open SharePoint **Central Admin** site.
2. Click **General Application Settings**.
3. Click **Farm Search Administration**.
4. Click **Search Service Application** link.
5. Click **Result Sources** under **Queries and Results**.
6. Click on **New Result Source**.
7. Enter **Bing** in **Name** box. Select **OpenSearch 1.0/1.1** in the **Protocol** section. In **Query Transform**, copy the following URL. Also copy this URL in the **Source Url**. In **Credentials information**, select **Anonymous**. Click **Save**.

   http://www.bing.com/search?q={searchTerms}&go=&form=QBLH&qs=n&format=rss

8. Go to SharePoint search center site that you created above. Add a new page and name it **BingSearch (bingsearch.aspx)**. Add another page and name it **Bing Search Results (bingsearchresults.aspx)**.

9. Add **Search** web part to the page. Edit web part properties. Select **Send queries to a custom results page URL** and **/searchcenter/Pages/bingsearchresults.aspx**. Remove the querystring appended automatically at the end of the URL (if you just replaced the url part otherwise if you overwrote the whole url then querystring will not be there). Click **OK**. Click **Save** button in the ribbon.

10. Go to **bingsearchresults.aspx** page and add **Search Results** web part. Edit web part properties. Click **Change Query** button. Select **Bing (Service)** from the **select a query** drop down. Click **OK**. Click **Save** button in the ribbon.

11. Go to **bingsearch.aspx** page and enter a search term and click search. You will see results from the web via Bing search engine.

You can set up search to retrieve results from Google, Youtube, Live, Twitter, and other sources. Follow the steps above and use relevant search query when setting up **Result Source**. You can stylize the results in the **Search Results** web part.

Search External Data Sources

Did you know with SharePoint you could search external data sources as well? You can search PHP sites, you can search line of business data, you can search files and folders on the server. Is there any other search tool that gives you much power and control? SharePoint comes with built-in mechanism and power to perform searches on all kinds of data sources. All you have to do is set it up. In this section, you will learn to configure some more search types, for example, searching PHP sites (both internal and external), searching folders, etc.

Search PHP Sites

As of this writing, PHP is the 6th most popular programming language (after Java, C, C++, C#, and Objective-C). A couple years back, it was at 4th position. I consider it to be a most popular web programming language. Till Jul 2007, 20,917,850 domains and 1,224,183 IP addresses had been registered for PHP sites. Searching sites created with such a popular technology with one of the world's best search technologies really makes sense. Especially in enterprise environments, where you are usually tasked with such requirements, this kind of combination does not seem illogical. Steps for setting up PHP search are as following:

1. Open SharePoint **Central Administration** site.

2. Click **General Application Settings**.

3. Click **Farm Search Administration**.

4. Click **Search Service Application**.

5. Click **Content Sources** under **Crawling** in the left side menu.

6. Click **New Content Source**.

7. In **Name** field, enter **PHP**.

8. In **Content Source Type**, select **Web Sites**.

9. In **Start Addresses**, enter the start address of the site that you want to crawl. If you have PHP setup in the local environment, you can enter the local address, for example, I have it set up locally so I entered http://127.0.0.1:8080. If you want to search an external PHP website, enter the address of that website. For example, I have a public website developed in PHP and I wanted to fetch results from that site therefore I entered the URL of my public site (which is http://walisystemsinc.com if you are curious to know). Once you have entered the starting address, you can move on to the other properties.

10. In **Crawl Settings**, you have three options:

    i) Only crawl within the server of each start address
    ii) Only crawl the first page of each start address.
    iii) Custom - specify page depth and server hops.

    You can choose the option that suits your needs. Selecting the first option will allow you to search the whole site starting at the start address you provided. Choose first option.

11. In **Crawl Schedules**, you can schedule a crawl. You can schedule both **Full Crawl** and **Incremental Crawl**. Click **Create schedule** link under **Full Crawl** drop down.

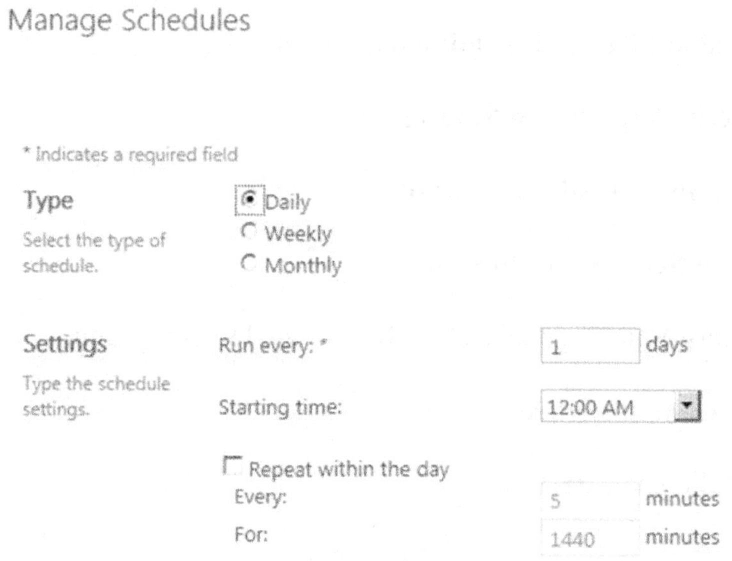

Figure 15-6: *Manage Schedules*

It's up to you if you want to schedule a crawl or not. You can do it manually whenever you want. In real world environment, you would setup an automatic crawl. If you want to schedule it, select the type and time and click **OK**.

12. You can select priority of this content source in **Content Source Priority**. By default, it is set to **Normal**. You can change it to **High**.

13. Click **OK**.

14. Right-click **PHP** content source and select **Start Full Crawl**. It will take a few minutes for the crawling to complete depending on the number of pages in the site. You can click the **Refresh** link to see updated status during the crawling process.

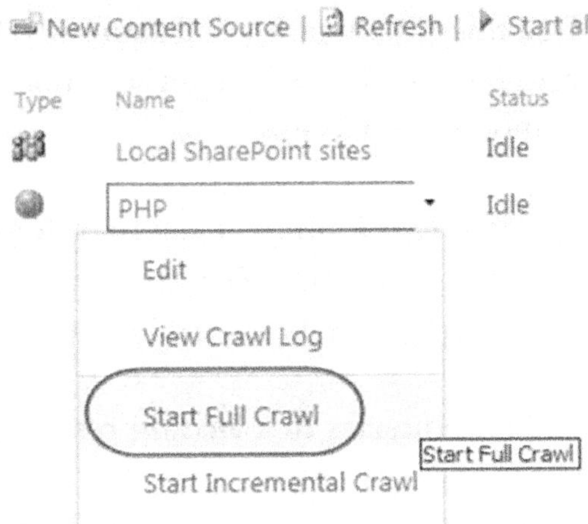

Figure 15-7: *Start Full Crawl*

15. Click **File Types** under **Crawling**. Make sure PHP is listed as one the file name extensions. If it's not there, click **New File Type** link at the top and enter **php** in the File extension field. Click **OK**.

16. Next, create a new **Result Source** for **PHP** search. Click **Result Sources** under heading **Queries and Results**. Creating new result source will let you apply filters on the search results.

17. Click **New Result Source**.

18. Enter **PHP Search** in **Title**. Enter **Search PHP Sites** in **Description**. In **Protocol**, select **Local SharePoint** (which is the default selection anyway). In **Type**, select **SharePoint Search Results**. If you want to apply filters to the search results, click **Launch Query Builder** and add filters otherwise move ahead. Keep **Default Authentication** selected in **Credentials Information**. Click **Save** to save the settings.

15. You are now ready to test the PHP searching. Before that, make sure the content has been crawled otherwise no results will appear in the search. Go to the SharePoint site. Go to **Site Settings > Site Contents**. Go to **Pages** library. If you don't see **Pages** library that's because you have not enabled **SharePoint Server Publishing** feature in **Site Settings**. To enable that feature in site settings, first you have to enable **SharePoint Server Publishing infrastructure** feature in site collection features.

16. Select **FILES** tab. Click **New Document** button. Enter **Custom PHP Search** in **Title**. Enter **phpsearch** in **URL Name**. Default selection in **Page Layout** is **(Article Page) Body only**. Click **Create**.

17. Add search web part to the page. Click **EDIT** from the top right corner. Click **INSERT** tab. Click **Web Part** button. Select **Search** from **Categories**. Select **Search Box** from the **Parts**. Click **Add** button.

18. Go to web part properties. Select **Send queries to a custom results page URL**. In **Results page URL**, enter **/searchcenter/Pages/phpsearchresults.aspx**. Replace **searchcenter** in the URL with your actual search site name. Make sure there is no querystring at the end of the URL. Click **OK**. This page has not been created yet. You will create this page now.

19. Create another page following the instructions given above. Name it **PHP Search Results** and enter **phpsearchresults.aspx** in the **URL**.

19. Go to the new page. Edit it and add a web part following the instructions given above. This time add **Search Results** web part. After adding the web part, edit its properties. Click **Change Query** button. In the **Select a query** drop down, select the **result source** (PHP (Service)) you created in the previous steps. In the **Keyword filter**, select **Query from the search box**. Click **OK**. Click **Save** button in the ribbon to save the page.

20. Go to the PHP search page and enter a search term and hit search. You will see results from the PHP site.

Figure 15-8: *PHP Search Results*

Searching Line of Business Data

Search is one of the top selling points for Microsoft and SharePoint. As I mentioned previously, SharePoint has one of the best search technologies in the world. No doubt, several big enterprises decide to buy SharePoint because of its excellent search capabilities. These companies know the importance of good quality data. Data is the basis for all research endeavors. Useful data helps stakeholders in their decision making. Collecting good quality data plays a vital role in supplying objective information for analytical understanding of the decision making problems and hence solutions can be obtained for such problems. Big companies value good quality data because making decisions on the basis of poor quality data is risky and may lead to disastrous results. Only good quality data can be used to generate information which helps stakeholders in solving their analytical problems. SharePoint allows you to setup search so that you can retrieve data from any kind of data source. You can get data from SharePoint sites, external sites, and even line of business applications. There are two parts to setting up line of business search. First you need to setup an application that will fetch data from the line of business application. This is done using Business Connectivity Services. Then you setup search to surface data from the external content types. You can learn more about Business Connectivity Services and External Content Types in the chapter that just focuses on this topic. I am going to skip that information in this chapter. If you want to explore this topic further, I would suggest you go back and read the dedicated chapter on this topic first and then come back to setup search.

You have seen how you can setup search in SharePoint to search external websites or non-SharePoint websites in your local environment. In this section, you will learn to setup search that will surface data from line of business applications. Most enterprise-scale web applications store their data in databases. If you have setup external content types, then you are ready to setup custom search to dig these content sources.

1. Open SharePoint **Central Administration** site.
2. Click **General Application Settings**.
3. Click **Farm Search Administration**.
4. Click **Search Service Application**.
5. Click **Content Sources** under **Crawling** in the left side menu.
6. Click **New Content Source**.
7. In **Name** field, enter **BCSSearch**. BCS stands for Business Connectivity Services.
8. In **Content Source Type**, select **Line of Business Data**.
9. In **External Data Source** select the business data connectivity service application from the drop down. If you have multiple search applications, you will see all of them listed here. If you only have the default service application then that will be selected by default.
10. There are two options for the crawling. One is to crawl all external data sources in the Business Data Connectivity Service Application and the other is to crawl selected external data source. You may have an inclination to select the first option but there is a problem with the first option. Crawl results in the following error:

Directory links across partitions are now allowed.

As of this time, there is no known solution to the problem except to crawl selected data sources. So, choose the second option **Crawl selected external data source** and check the sources that you want to search.

11. Click **Create Schedule** under **Full Crawl** (in **Crawl Schedules** section) and setup a schedule and click OK.

12. Click **Ok**.

13. On **Manage Content Sources** page, right-click **BCSSearch** content source and select **Start Full Crawl**. It will take some time for the crawling to complete depending on the amount of data in the external content sources. Once the crawl is complete, right-click **BCSSearch** and select **View Crawl Log**. Look at the **Errors** column. If the number is greater than 0, click the number to view the errors. Column **Successes** show the number of successful crawls. When you click the number under **Errors**, you see the error count and the message. Click the error count again to view the detailed error message.

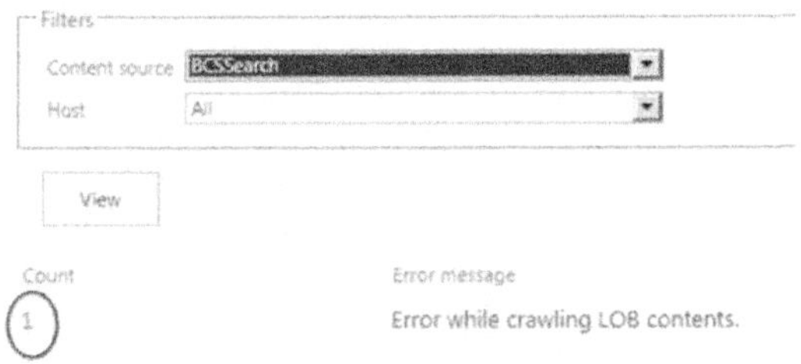

Figure 15-9: *Click error count to view detailed message*

The most common error that you get during the crawl process is the **Access Denied** error.

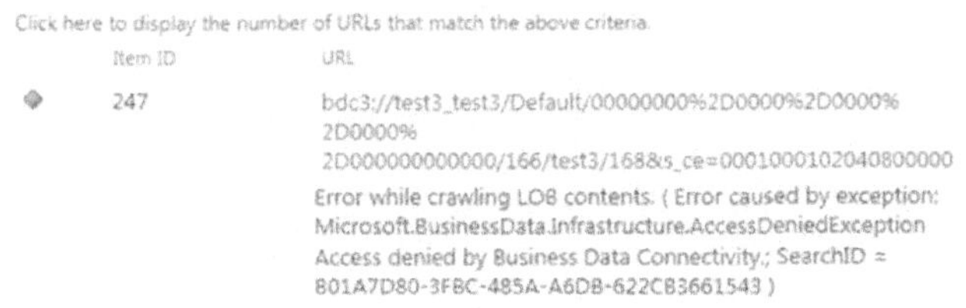

Figure 15-10: *Error while crawling LOB contents*

This error occurs if the service account does not have permissions on one or all of the external content type objects. To see how to resolve this issue, read the chapter on Business Connectivity Services. This error and its solution have been discussed in detail in that chapter. Once you have resolved the issue, start the crawl again.

14. Go to SharePoint site and create search pages to test the line of business search. I am not going to repeat the steps again. Please read previous sections if you are not sure how to create search and search results pages. Once you have created the search results page and added the **Search Results** web part, edit the web part properties and click **Change query** button.

15. Select **Local SharePoint Results (System)** in the query drop down. By default, **Query text** box has **{SearchBoxQuery}**. After **{SearchBoxQuery}** enter **ContentSource:BCSSearch** in the **Query text** box. This will return results only from the content source that you specifically created for searching line of business data. Click **OK** to save the changes.

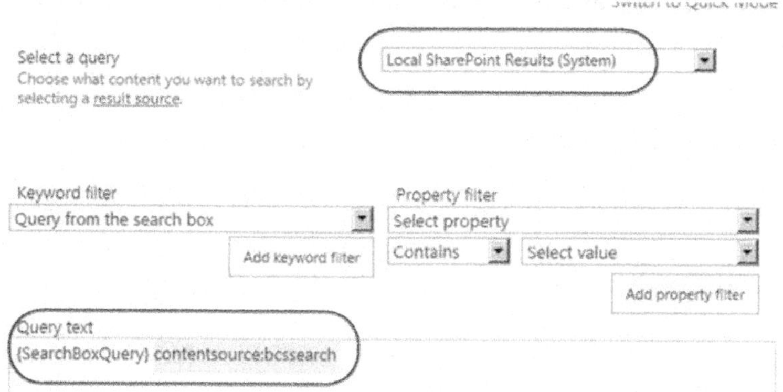

Figure 15-11: *Limit Search to BCS Content Source*

16. Go to the search page and enter a search term and hit search. You will see results from the external content types.

Figure 15-12: *BCS Search Results*

Searching File Shares

You can also search files shares for content and information using SharePoint search. To setup file share search, follow these steps:

1. Open SharePoint **Central Administration** site.

2. Click **General Application Settings**.

3. Click **Farm Search Administration**.

4. Click **Search Service Application**.

5. Click **Content Sources** under **Crawling** in the left side menu.

6. Click **New Content Source**.

7. In **Name** field, enter **FSSearch**. In **Content Source Type** select **File Shares**. In **Start Addresses** type the file share that you want to search. In **Crawl Settings**, you have two options. You can crawl the main folder and all subfolders or you can crawl only the main folder. Choose your preferred option. In **Crawl Schedules**, click **Create Schedule** under **Full Crawl** and setup a schedule. Click **OK**.

8. On **Manage Content Sources** page, right-click **FSSearch** content source and select **Start Full Crawl**. It will take some time for the crawling to complete depending on the amount of data in the file share.

9. Once the crawl is complete, go to SharePoint site and create search and search results pages. I will not repeat the steps in this section. Once you have created and configured the search pages, try out the search. Just remember one thing, you have to add **contentsource:fssearch** in the **Query text** box in the search results web part. FSSearch is the name of the content source you created in the previous step. By adding this content source in the **Query text** you limit the results to this content source only. See previous section for details if you have any confusion. You will see results for your searched term on a SharePoint page. This functionality is really useful for companies that hesitate in bringing all data from file shares into SharePoint. Some companies avoid it because of the storage costs and other limitations. SharePoint allows you to search files that are sitting outside SharePoint but still can be indexed and surfaced within SharePoint sites.

Readme.txt
This is a **SharePoint** File Search Test
sp2013/temp/Readme.txt

How to use SharePoint 2013.txt
Learn how to use **SharePoint** 2013 in book named A Practical Guide To **SharePoint**
sp2013/temp/How to use **SharePoint** 2013.txt

Figure 15-13: *File share search results*

Managed Metadata in SharePoint 2013

Since we are discussing search in SharePoint, I would like to briefly touch upon the topic of managed metadata. It's a big topic. Metadata has gained popularity over the last few years. It greatly enhances search functionality. Managed metadata is a hierarchical collection of centrally managed terms that you can define and then use as attributes for items in SharePoint. Metadata is information about information. You can add extra tags to the information to identify it easily. It's also called as labelling. You can label your documents or items and items can be searched by users using these labels or tags. These labels or tags are metadata. For example, if you have a contract document stored in SharePoint, you can label it using keywords like contract, case number, attorney name, law practice name, etc. Metadata is managed centrally. It's organized in a way that makes sense in your business. It is important to understand some key terminology.

Taxonomy

Taxonomy is a formal classification system. Taxonomy groups the words, labels, and terms that describe something, and then arranges the groups into a hierarchy. Taxonomy plays an important part in the management of information. Each organization defines different taxonomies for their different systems or departments. Some organizations prefer a single taxonomy for their business operations. For example, taxonomy for the HR department lists all terms related to HR operations. Taxonomy for finance department lists down all commonly used terms in finance.

Folksonomy

A folksonomy is an informal classification system. It evolves gradually as users collaborate on words used in a site. The concept is same as bookmarking. The tag clouds you see on blogs are an example of folksonomy.

Term Set

A term set is a group of related terms. Term sets can have different scope, depending on where you create the term set. Local term sets are created within the content of a site collection and are available only to the users of that site collection. Global term sets are available to all users across the SharePoint farm. Global term sets are available to sites that subscribe to managed metadata service.

Terms

A term is a specific word or keyword that you associate with an item. It is a single item in a term set. A term is unique because it has a unique ID and it can have many text labels. For example, in a multilingual site, the term can have different labels for different languages. Managed terms are pre-defined. Administrator defines these terms. Enterprise keywords are terms that site user adds to items. The collection of enterprise keywords is known as the keywords set.

Group

Group is a collection of term sets. Only users having contribute permissions can manage term sets within a group.

Term Store Management Tool

The term store management tool is the tool that is used to define term sets and terms. It is used by the administrators and is available in the central administration site under Search.

Managed Metadata column

This is special kind of column that you can add to lists. It is used to attach metadata to the items. This is an auto-populate column. As soon as user starts typing, SharePoint displays lists of terms available in the term set. User has to select a term set when adding this column to the list. User can then select the terms that are relevant to their item. This is also called as tagging.

Enterprise keywords column

This is a column that users can add to content types, lists, or libraries to enable users to tag items. It is a multi-value column. Users can add new terms in this column or select existing ones.

Social Tags

Social tags are tags that site users can apply to content to help them categorize information. Social tagging is useful because it helps site users to improve the discoverability of information on a site. Social tag consists of three parts: User identity, item URL, term. Term is stored in a term stored and is public information. User identity can be made private if user wants to stay anonymous. If user enters new keywords when tagging information, these keywords are stored in the keywords term set. Social tags support folksonomy-based tagging.

Metadata Advantage

The metadata allows users to tag items consistently. Centrally stored term stores are used by all users, therefore, users don't have to enter their own version of keywords. Common keywords available in the term store make it easier for the users to tag items. Biggest advantage of metadata is information discoverability. It is easier to find information. Information management has always been a problem especially in big companies and organizations where the volume of data is huge. It is important to convert data into information because information provides value. Discoverability is the first step towards achieving that goal. If information is stored somewhere and is not easily discoverable, that information is useless because it is not being used. Metadata is especially useful in the business environments where information is stored in different formats and across different sites. Using metadata, searching this scattered information become easy.

Metadata navigation

Metadata navigation enables users to create view dynamically, based on specific metadata fields.

Planning Metadata

Careful planning is very important before managed metadata service can be used. This is part of governance and is considered best practice. Each organization should have a taxonomy and based on taxonomy administrators can define global term sets. Planning metadata in advance helps administrators run operations smoothly. Problems start to appear when the size of data starts growing with the passage of time. As the number of users grows, and their usage of sites grows, data also grows quickly. With more documents stored in sites, and if users have adopted metadata, the size of metadata database also grows quickly. It is important to plan out things in advance otherwise managing and rectifying things at a later stage becomes problematic and time consuming.

Metadata Management

In this section, you will learn how to manage metadata and see a few practical examples.

1. Go to Central Administration site. Go to **General Application Settings**.

2. Click **Farm Search Administration** under **Search**.

3. Click **Search Service Application** link.

4. Under **Queries and Results**, click **Search Dictionaries**. **Term Store Management Tool** will open. Here you can define new groups, term sets and terms.

5. Expand **Managed Metadata Service** and select **New Group**. Enter **Custom** as the group name and hit Enter. Enter description, add group managers and contributors and click **Save**.

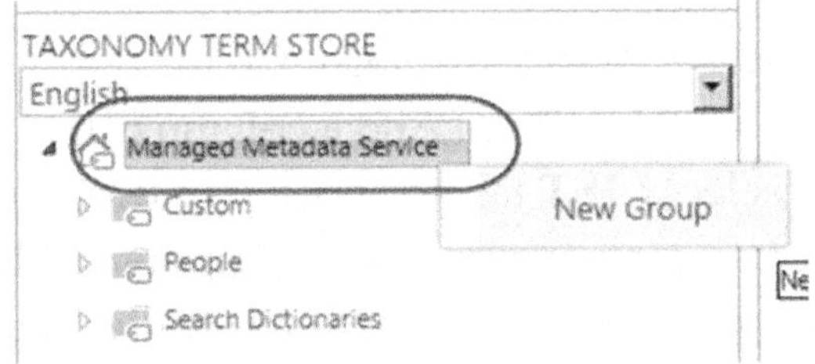

Figure 15-14: *Managed Metadata Service*

6. Click the new group and open drop down and select **New Term Set**. Enter **CustomTags** as the name of the term set and hit **Enter**. Add description, owner, contact, stakeholders and click **Save**. In the **Submission Policy**, you can choose to keep term set open or closed. When a term set is closed, only metadata managers can add terms to this term set. When it is open, users can

add terms from a tagging application.

7. Click on the term store and select **Create Term**. Enter a term and hit Enter. Add as many terms you like.

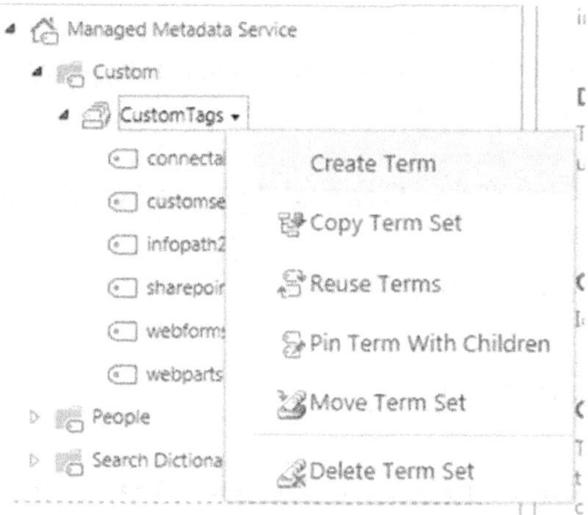

Figure 15-15: *Create Term*

You can see the other available options in the figure above. You can copy an existing term ste, you can reuse terms, you can pin term with children and you can move term set. **Pin Term With Children** means that you can pin this term with another term store.

8. You have now defined a term set. Let's use it in a list. Go to your SharePoint site and create a custom list. Call it **Metadata**. Add a column called **ItemType**. To create column, select **LIST** tab and click **Create Column** button. Enter **Metadata** in the column name and select **Managed Metadata** as the type. Check **Allow multipe values**. In **Term Set Settings**, select the newly created term set. In **Allow Fill-in**, choose **Yes**. Click **OK**.

9. Enter a new item in the list. Enter something in **Title**. Click in **ItemType** field and start typing. SharePoint will show you suggestions from the term set. Once you have attached metadata to the item you can search for the item using metadata terms in search.

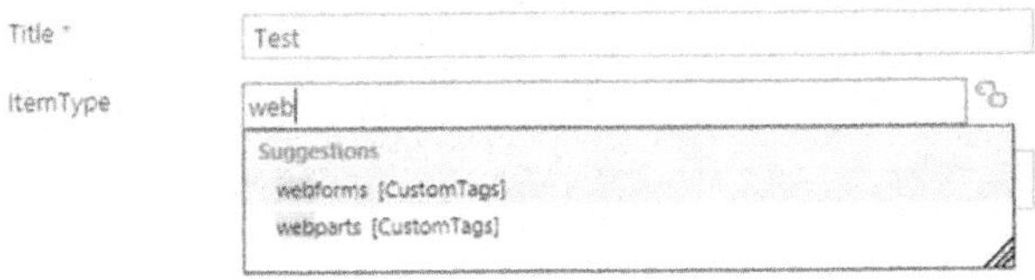

Figure 15-16: *Metadata in action*

Metadata Navigation and Filtering

Metadata navigation and filtering can be enabled in lists and libraries. Using metadata navigation makes it easier for users to find content in large lists and libraries. Users can easily filter and find content by using a navigation tree. It is separate from the search functionality that is available in the list. When setting up metadata navigation, you can determine which columsn from the library appear in the tree, and you can also specify key filters.

1. To enable metadata navigation, first you have to activate **Metadata Navigation and Filtering** feature in site features. Once you have activated this feature, you will see an additional setting in the list settings. Go to list settings, under **General Settings**, you will see **Metadata navigation settings**. Click this link.

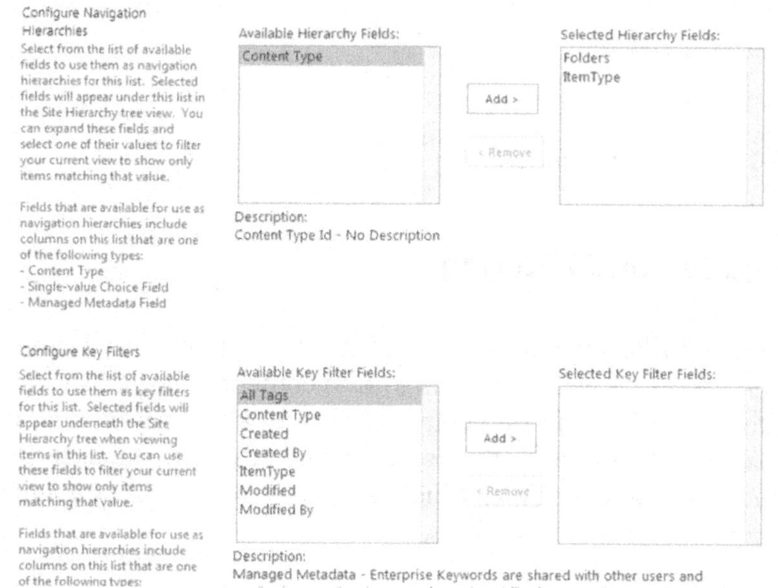

Figure 15-17: *Metadata Navigation Settings*

Add **ItemType**, the new column you created, from **Available Hierarchy Fields** to **Selected Hierarchy Fields**. You can also add filter fields that you want to appear in the tree navigation.

2. This is how navigation tree looks:

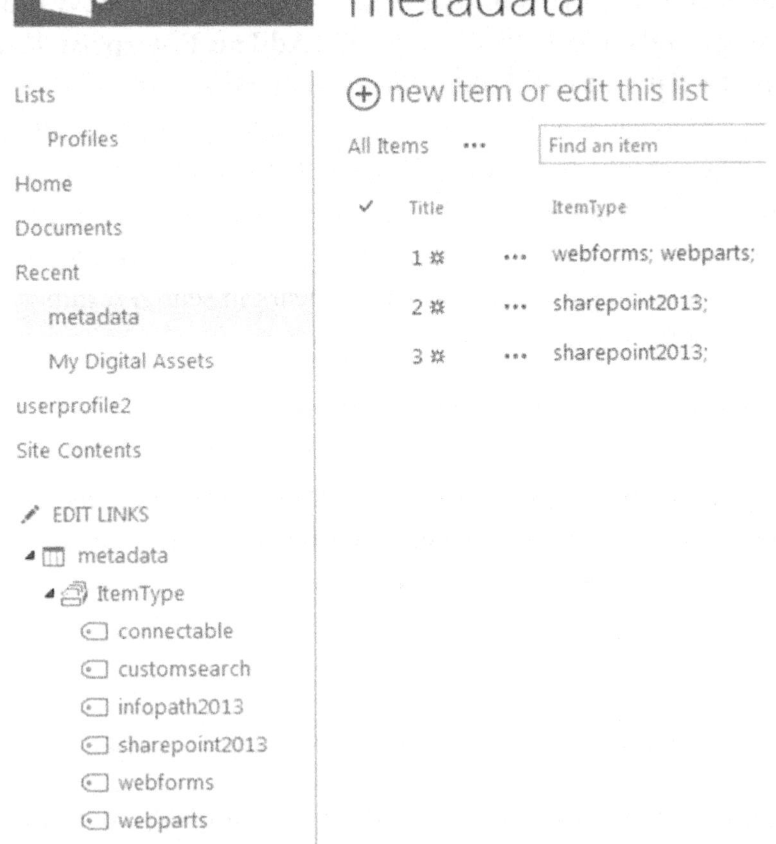

Figure 15-18: *Metadata navigation and filtering*

Click on any term in the tree to filter the results, you will only see items related to that term. The tree displays a hierarchy of managed terms. This lets users create dynamic displays of content. Users can refine this more by using the key filters. You may have noticed another option on the settings page that lets you create index. By default, SharePoint performs automatic index creation. Indexes are created automatically for the fields that you add to the navigation tree. You can disable automatic indexing and configure the indexes manually. If a user loads a view that exceeds the threshold, then metadata navigation and filtering creates a modified version of the original user query called a fallback query. The fallback query runs against a portion of the list instead of the complete list. As a result, it displays a partial set of the items requested.

3. In list settings, there is another option called **Enterprise Metadata and Keywords Settings** under **Permissions and Management**. If enabled, it lets

users enter their own keywords. All keywords are stored in keywords set. Go to list settings, and click on this link. Check **Add an Enterprise Keywords column to the list and enable keyword synchronization**. Now go back to the list and add a new item. You will see two fields, one the custom metadata column and another for keywords. If term set is closed, users cannot add their own terms but they can still add keywords using the keywords field and it helps them collaborate and share information with each other because this information then appears in search results when someone performs a search using these keywords.

Metadata Management in SharePoint Online

1. Now, you will enable metadata management in SharePoint Online. Log on to SharePoint Online. Go to **Site collection features** and activate feature **SharePoint Server Standard Site Collection features**.

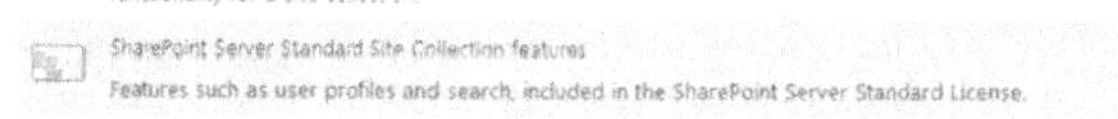

Figure 15-19: *SharePoint Server Standard Site Collection features*

2. Go to **Manage site features** and activate **SharePoint Server Publishing** feature.

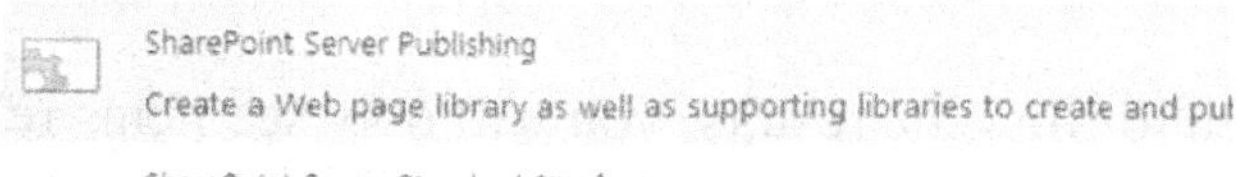

Figure 15-20: *SharePoint Server Publishing*

3. After activating the feature, you will see a link called **Navigation** in **Site Settings**.

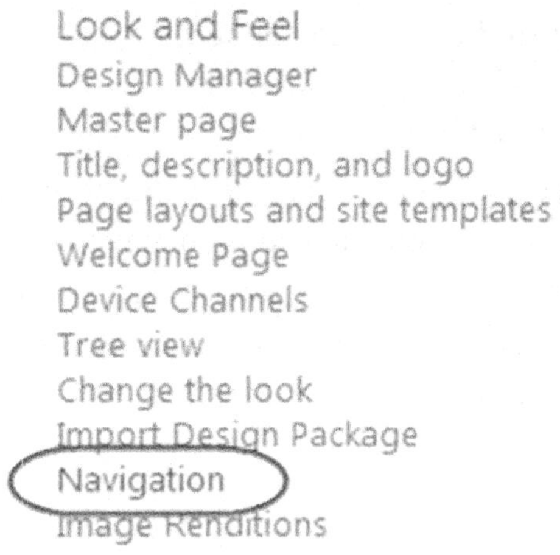

Look and Feel
Design Manager
Master page
Title, description, and logo
Page layouts and site templates
Welcome Page
Device Channels
Tree view
Change the look
Import Design Package
Navigation
Image Renditions

Figure 15-21: *Navigation*

4. Click this link to go to **Navigation Settings**. In **Global Navigation**, select **Managed Navigation** option.

Site Settings › Navigation Settings

Figure 15-22: *Managed Navigation*

5. In **Current Navigation**, select **Managed Navigation** (the second option).

6. In **Managed Navigation: Term Set** section, click **Create Term Set** button. This will add a new term set. You will see message "Successfully created term set 'xyz'. If your site name is ABC Inc., the term set group will be called as **ABC Inc. Team Site Navigation**. You can create more term sets if you want by clicking the **Create Term Set** button. All term sets will appear under the same group.

7. Under the **Create Term Set** button, there will be a link **Term Store Managemen Tool**. Click it to open the tool and add terms.

8. Under **Site Collection – Your site URL** link, you will see your new term set. Open its menu and select **Create Term**. Enter term and hit Enter. Add as many terms as you like. Remember, you can group managers and contributors in the settings on the right side. You can also enter root URLs of site collections that will have read only access to the contents of term sets in this group in the **Site Collection Access**.

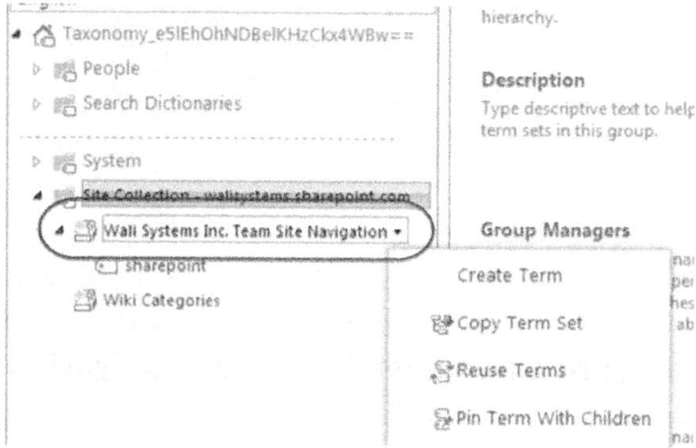

Figure 15-23: *Add new terms*

9. After you have added term, click on the term. It will open term's settings on the right side. You can change the default label of the term. You can also add other labels for the term. You can enable or disable tagging for this term by checking/unchecking **Available for Tagging** option.

10. You can enter children terms as well. For example, select a term and then add terms under it. You can create a hierarchy of terms.

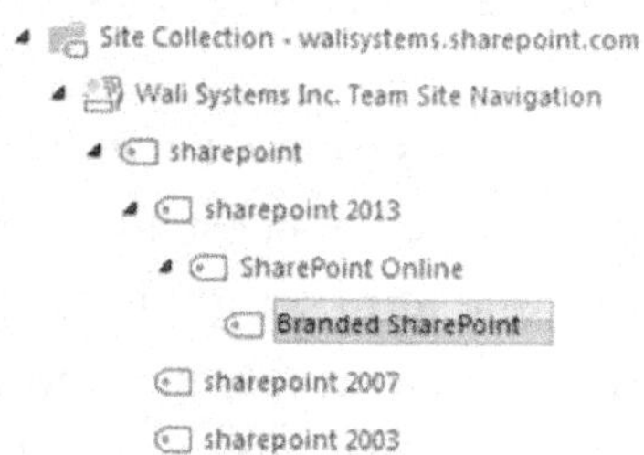

Figure: 15-24: *Terms hierarchy*

11. Click on the term set again. On the right side, you will see general properties of the term set. There are several other tabs that you can use to definre more properties for the term set. Select **INTENDED USE** tab. You can choose the option of showing or hiding this term set on this page. You can also choose if you want to use this term set for site navigation. You can define custom sort for your terms by going to **CUSTOM SORT** tab. Go to **TERM-DRIVEN PAGES** tab.

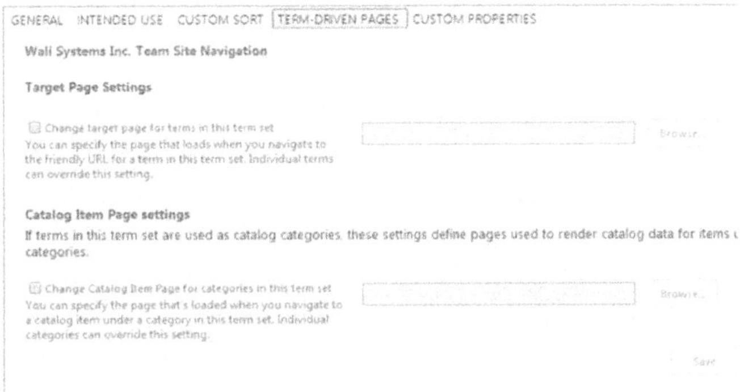

Figure 15-25: *Term-drive pages*

Target Page Settings allow you to change target page for terms in this term set. You can specify the page that loads when you navigate to the friendly URL for a term in this term set. **Catalog Item Page settings** allow you to define pages used to render catalog data. The terms in this term set can be used as catalog categories. You can specify the page that's loaded when you navigate to a catalog item under a category in this term set.

CUSTOM PROPERTIES tab lets you define custom properties to store additional data about a term set. If you select a term and then go to **CUSTOM PROPERTIES** tab, you will see **Shared Properties** and **Local Properties**. Shared Properties are available on all reused or pinned instances of this term anywhere in the term store. Local Properties are only available for this term in this term set.

12. Go to main site and add a custom list or use an existing list to test metadata terms that you added in the term set. You can use default **Documents** library in the main site to test the metadata terms. Go to **Documents** library. Click **LIBRARY** tab. Click **Create Column** button in the ribbon. Enter **My Terms** in the **Column name**. Select **Managed Metadata** as the column type. In **Term Set Settings**, select the term set you created above. Click **OK**.

13. Add a new document in the library. Go to document properties and select **Edit Properties**. On **Edit Properties** page, you will see metadata column **My Terms**. Start typing something in the field, you will see the terms related to the word being entered in the suggestions. Select terms and click **Save**.

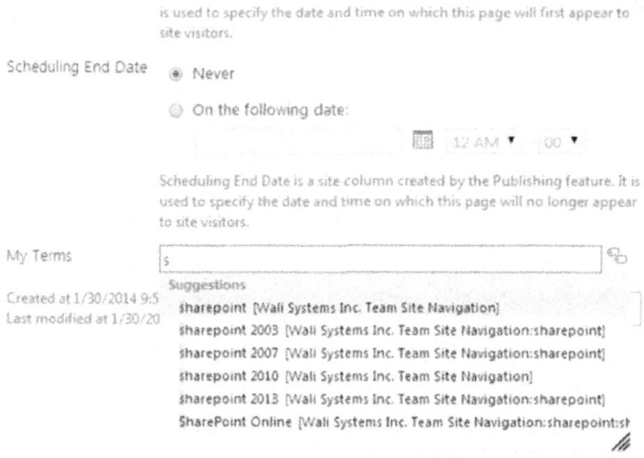

Figure 15-26: *Custom terms*

Summary

SharePoint enterprise search is regarded as one of the world's best search technologies. SharePoint 2013 has gone one step farther and added new enhancements to this world-class technology. SharePoint search provides two approaches for processing queries to return search results. One is content crawling and the other is federated search. You can search PHP sites, you can search line of business data, you can search files and folders on the server. You can enhance search by enabling managed metadata and creating term sets.

Index

Symbols

A

B

C

D

E

F

G

H

I

J

K

L

M

N

O

P

Q

R

S

T

U

V

W

X

Y

Z

CPSIA information can be obtained
at www.ICGtesting.com
Printed in the USA
LVOW05s0326011017
550696LV00006B/769/P